STUDY GUIDE

FOR

STEWART'S

MULTIVARIABLE CALCULUS

FIFTH EDITION

Richard St. Andre
Central Michigan University

THOMSON

BROOKS/COLE

Australia • Canada • Mexico • Singapore • Spain • United Kingdom • United States

Printed in the United States of America
3 4 5 6 7 07 06 05

Printer: Globus

ISBN 0-534-39358-6

For more information about our products,
contact us at:
Thomson Learning Academic Resource Center
1-800-423-0563

For permission to use material from this text,
contact us by:
Phone: 1-800-730-2214
Fax: 1-800-730-2215
Web: http://www.thomsonrights.com

Brooks / Cole—Thomson Learning
10 Davis Drive
Belmont, CA 94002-3098
USA

Asia
Thomson Learning
5 Shenton Way #01-01
UIC Building
Singapore 068808

Australia
Nelson Thomson Learning
102 Dodds Street
South Street
South Melbourne, Victoria 3205
Australia

Canada
Nelson Thomson Learning
1120 Birchmount Road
Toronto, Ontario M1K 5G4
Canada

Europe/Middle East/South Africa
Thomson Learning
High Holborn House
50/51 Bedford Row
London WC1R 4LR
United Kingdom

Latin America
Thomson Learning
Seneca, 53
Colonia Polanco
11560 Mexico D.F.
Mexico

Spain
Paraninfo Thomson Learning
Calle/Magallanes, 25
28015 Madrid, Spain

Preface

This *Study Guide* is designed to supplement the multivariable calculus chapters of *Calculus*, 5th edition, by James Stewart. It may also be used with *Calculus, Early Transcendentals*, 5th edition and *Multivariable Calculus*, 5th edition.

Your text is well written in a very complete and patient style. This Study Guide is not intended to replace it. You should read the relevant sections of the text and work problems, lots of problems. Calculus is learned by doing; it is not a spectator sport.

This *Study Guide* captures the main points and formulas of each section and provides short, concise questions that will help you understand the essential concepts. Every question has an explained answer. The two-column format allows you to cover the answer to check your solution. Working in this fashion leads to greater success than simply perusing the solutions. Students have found this *Study Guide* helpful for reviewing for examinations.

Technology can be a tool to help understand calculus concepts by drawing intricate graphs, solving or approximating difficult equations, doing numerically intense work, and performing symbolic manipulations. We have included additional questions in sections called "Technology Plus" at the end of each chapter. For these questions you should use a graphing calculator or computer with a computer algebra system software program.

As a quick check of your understanding of a section you should work the page of On Your Own questions located toward the back of the *Study Guide*. These are all multiple choice type questions — the kind that you might see on an exam in a large-sized calculus class. You are "on your own" in the sense that an answer, but no solution, is provided for each question.

I hope that you find this *Study Guide* helpful in understanding the concepts and solving the exercises in *Calculus*, 5th edition.

Richard St. Andre

Table of Contents

Please cut page down center line.
Use the half page to cover the right
column while you work in the left.

Chapter 11 — Parametric Equations and Polar Coordinates

Cartoons courtesy of Sidney Harris. Used by permission.

Section 11.1 Curves Defined by Parametric Equations

As a point (x, y) moves along a curve over a time interval, the coordinates x and y may each be described as a function of a time variable t. The notion of describing a curve with a pair of parametric equations (t is the parameter) is covered in this section. We will see that graphs of ordinary functions ($y = f(x)$) are one of many types of curves that may be defined parametrically.

Concepts to Master

Parameter; Parametric equations, graphs of curves defined parametrically; Elimination of the parameter

Summary and Focus Questions

Page 687 (ET Page 651)*

A set of *parametric equations* has the form

$$x = f(t)$$
$$y = g(t),$$

where f and g are functions of a third variable t, called a *parameter*. Each value of t determines a point (x, y) in the plane. The collection of all such points is a *curve*.

A curve may be described by several different pairs of equations (each version called a "parametric curve"). Here are two sets of parametric equations for the quarter of the unit circle in the first quadrant:

$$x = t,$$
$$y = \sqrt{1 - t^2}$$
$$\text{for } t \in [0, 1]$$

$$x = \cos t,$$
$$y = \sin t$$
$$\text{for } t \in \left[0, \frac{\pi}{2}\right]$$

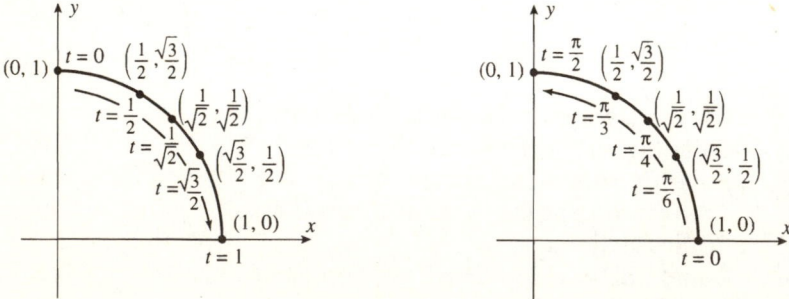

Each has the same graph (set of points). The one on the left is traversed clockwise as t increases; the one on the right is traversed counterclockwise.

*Remember, when using the Early Transcendentals book, use the page in parentheses!

Some parametric equations may be combined algebraically to a single equation not involving the parameter t; this process is called *eliminating the parameter*. The method to do so depends greatly on the nature of the parametric equations involved. Sometimes identities need to be employed, especially if trigonometric functions are involved; sometimes you can solve for t in one parametric equation and substitute the result in the other equation.

Example: Eliminate the parameter in $x = 2 + 3t$, $y = t^2 - 2t + 1$.

Solve $x = 2 + 3t$ for t to get $t = \frac{1}{3}(x - 2)$. Substitute this in the other equation:

$$y = \left[\tfrac{1}{3}(x - 2)\right]^2 - 2\left[\tfrac{1}{3}(x - 2)\right] + 1.$$

$$y = \tfrac{1}{9}x^2 - \tfrac{10}{9}x + \tfrac{25}{9}.$$

Example: Eliminate the parameter in $x = \sin t$, $y = \cot^2 t$.

$y = \cot^2 t = \dfrac{\cos^2 t}{\sin^2 t} = \dfrac{1 - \sin^2 t}{\sin^2 t}$. Thus $y = \dfrac{1 - x^2}{x^2}$.

We note from $x = \sin t$, $-1 \le x \le 1$ and $x \ne 0$ (since $\cot t$ is undefined when $\sin t = 0$.)

1) Sketch a graph of the curve given by
$x = \sqrt{t}$, $y = t + 2$.

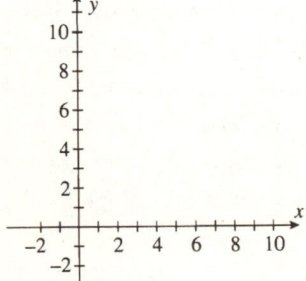

Compute some values and plot points:

t	0	1	4	9
x	0	1	2	3
y	2	3	6	11

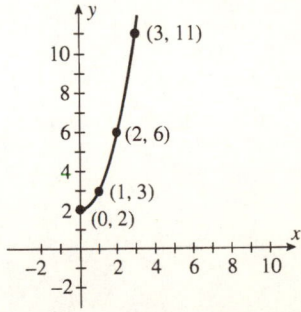

We see this is half a parabola since $x = \sqrt{t}$ implies $x^2 = t$.
Thus $y = x^2 + 2$, $x \ge 0$.

2) Eliminate the parameter in each

    a) $x = e^t, y = t^2$

Solve for t in terms of y:
$y = t^2, \sqrt{y} = t$.
Thus $x = e^{\sqrt{y}}, x > 0$.
A solution may also be obtained by solving for t in terms of x:
$x = e^t, t = \ln x, y = (\ln x)^2, x > 0$.

    b) $x = 1 + \cos t, y = \sin^2 t$

$x = 1 + \cos t, x - 1 = \cos t$
$\cos^2 t = (x - 1)^2$
Since $\sin^2 t = y$, and $\cos^2 t + \sin^2 t = 1$
we have $(x - 1)^2 + y = 1$.
Thus $y = 1 - (x - 1)^2$.
Since $x = 1 + \cos t, 0 \leq x \leq 2$.

    c) $x = 2 \sec t, y = 3 \tan t$

$\sec t = \frac{x}{2}$ and $\tan t = \frac{y}{3}$.
Hence the identity $\tan^2 t + 1 = \sec^2 t$
becomes $\left(\frac{y}{3}\right)^2 + 1 = \left(\frac{x}{2}\right)^2$ or $\frac{x^2}{4} - \frac{y^2}{9} = 1$.

3) Describe the graph of $x = a + bt$, $y = c + dt$ where $a, b, c,$ and d are constants.

If both b and d are zero, this is the single point (a, c). If $b = 0$ and $d \neq 0$ this is a vertical line through (a, c). If $b \neq 0$, then $t = \frac{x - a}{b}$ and $y = c + d\frac{(x - a)}{b}$. Thus $y = \frac{d}{b}x + c - \frac{da}{b}$, so the graph is a line through (a, c) with slope $\frac{d}{b}$.

4) Does the pair
$$x = e^t$$
$$y = e^{2t}, \quad -\infty < t < \infty$$
represent the parabola $y = x^2$?

Since $x = e^t$, $x > 0$. Thus the pair represent that portion of $y = x^2$ to the right of the y-axis.

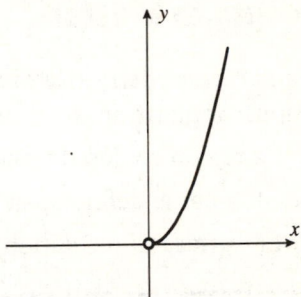

5) Use a graphing calculator to sketch
$$x = t^2 + 3t$$
$$y = 2 - t^3$$

t	x	y
-4	4	66
-3	0	29
-2	-2	10
-1	-2	3
0	0	2
1	4	1
2	10	-6
3	18	-25
4	28	-62

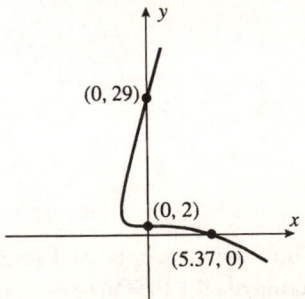

Section 11.2   Calculus with Parametric Curves

This section describes several familiar calculus concepts for curves defined parametrically: slope of a tangent line to a curve, curve sketching, length of a curve, areas of regions enclosed by curves and volumes of surfaces and solids of revolution.

Concepts to Master

A. First and second derivative of a function defined by a pair of parametric equations; applications to graphs of curves defined parametrically

B. Area of a region enclosed by curves defined parametrically

C. Length of a curve defined parametrically

D. Area of a surface of revolution with curve defined parametrically

Page 696 (ET Page 660)

Summary and Focus Questions

A. The slope of the line tangent to a curve described by $x = f(t)$, $y = g(t)$ is the first derivative of y with respect to x and is given by

$$\frac{dy}{dx} = \frac{\frac{dy}{dt}}{\frac{dx}{dt}} \text{ if } \frac{dx}{dt} \neq 0.$$

The curve will have a horizontal tangent when $\frac{dy}{dt} = 0$ and will have a vertical tangent when $\frac{dx}{dt} = 0$ and $\frac{dy}{dt} \neq 0$.

1) For $x = t^3$, $y = t^2 - 2t$

a) Find $\frac{dy}{dx}$ and $\frac{d^2y}{dx^2}$.

$\frac{dx}{dt} = 3t^2$ and $\frac{dy}{dt} = 2t - 2$, so $\frac{dy}{dx} = \frac{2t - 2}{3t^2}$.

$\frac{d}{dt}\left(\frac{dy}{dx}\right) = \frac{3t^2(2) - (2t - 2)(6t)}{(3t^2)^2} = \frac{4 - 2t}{3t^3}$.

Thus $\frac{d^2y}{dx^2} = \frac{\frac{4 - 2t}{3t^3}}{3t^2} = \frac{4 - 2t}{9t^5}$.

b) Find the horizontal and vertical tangents for the curve.

$\frac{dy}{dx} = 0$ at $t = 1$ ($x = 1$, $y = -1$).
A horizontal tangent is at $(1, -1)$.
$\frac{dy}{dx}$ does not exist at $t = 0$ ($x = 0$, $y = 0$).
A vertical tangent is at $(0, 0)$.

c) Discuss the concavity of the curve.

$\dfrac{d^2y}{dx^2} > 0$ for $0 < t < 2$ $(0 < x < 8)$ and negative for $t < 0$ $(x < 0)$ and $t > 2$ $(x > 8)$. The curve is concave upward for $x \in (0, 8)$ and concave downward for $x \in (-\infty, 0)$ and $x \in (8, \infty)$.

d) Sketch the curve using the information above.

$\dfrac{d^2y}{dx^2} = \dfrac{4 - 2t}{9t^5} = 0$ at $t = 2$ $(x = 8, y = 0)$.
$(8, 0)$ and $(0, 0)$ are inflection points.

t	x	y
−4	−64	24
−3	−27	15
−2	−8	8
−1	−1	3
0	0	0
1	1	−1
2	8	0
3	27	3
4	64	8

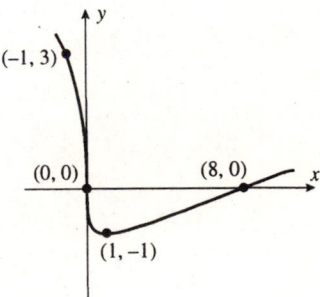

2) Find the equation of the tangent line to the curve $x = 4t^2 + 2t + 1$, $y = 7t + 2t^2$ at the point corresponding to $t = -1$.

At $t = -1$, $x = 3$ and $y = -5$.

$\dfrac{dx}{dt} = 8t + 2 = -6$ at $t = -1$.

$\dfrac{dy}{dt} = 7 + 4t = 3$ at $t = -1$.

Thus $\dfrac{dy}{dx} = \dfrac{3}{-6} = -\dfrac{1}{2}$. The tangent line is

$y + 5 = -\dfrac{1}{2}(x - 3)$.

Page
698
(ET Page
662)

B. Suppose the parametric equations $x = f(t), y = g(t), t \in [\alpha, \beta]$ define an integrable function $y = F(x) \geq 0$ over the interval $[a, b]$, where $a = f(\alpha)$, $b = f(\beta)$. The area under $y = F(x)$ is $\int_a^b y \, dx =$

$$\int_\alpha^\beta g(t) f'(t) \, dt \quad \text{if } (f(\alpha), g(\alpha)) \text{ is the left endpoint}$$

or

$$\int_\beta^\alpha g(t) f'(t) \, dt \quad \text{if } (f(\beta), g(\beta)) \text{ is the left endpoint.}$$

3) Find a definite integral for the area under the curve $x = t^2 + 1, y = e^t, 0 \leq t \leq 1$.

The area is $\int_0^1 e^t \, (2t) dt$

(by parts: $u = 2t$ and $dv = e^t \, dt$; $du = 2 \, dt, v = e^t$)

$$= 2te^t - \int e^t 2 \, dt = 2te^t - 2e^t \Big]_0^1$$
$$= (2e - 2e) - (0 - 2) = 2.$$

4) Find the area inside the loop of the curve given by
$$x = 9 - t^2$$
$$y = t^3 - 3t.$$

$$\frac{dy}{dx} = \frac{\dfrac{dy}{dt}}{\dfrac{dx}{dt}} = \frac{3t^2 - 3}{-2t}.$$

$\frac{dy}{dx} = 0$ at $t = 1, t = -1$. $\frac{dy}{dx}$ is not defined at $t = 0$. There are horizontal tangents at $(8, 2)$ (where $t = -1$) and $(8, -2)$ (where $t = 1$). There is a vertical tangent at $(9, 0)$ (where $t = 0$).

t	x	y
-3	0	-18
-2	5	-2
-1	8	2
0	9	0
1	8	-2
2	5	2
3	0	18

The graph is

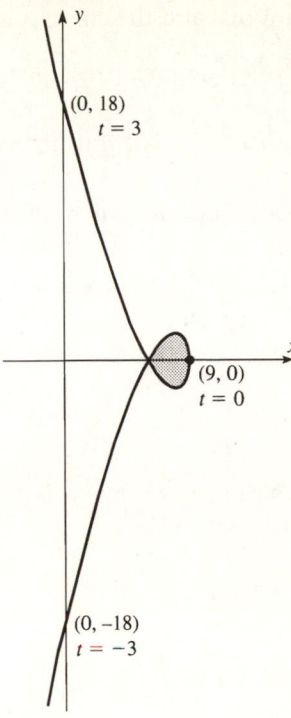

The curve crosses itself (to form a loop) at $y = 0$.

$t^3 - 3t = 0$

$t(t^2 - 3) = 0$

$t = 0$, $t = \sqrt{3}$, $t = -\sqrt{3}$.

The area of the loop is twice the area under the top portion which corresponds to $t \in [-\sqrt{3}, 0]$.

The area is $2 \displaystyle\int_{-\sqrt{3}}^{0} (t^3 - 3t)(-2t)\, dt$

$= 2 \displaystyle\int_{-\sqrt{3}}^{0} (-2t^4 + 6t^2)\, dt$

$= 2 \left(-\tfrac{2}{5}t^5 + 2t^3 \right) \Big]_{-\sqrt{3}}^{0}$

$= \dfrac{24\sqrt{3}}{5}.$

Page
699
(ET Page
663)

C. The length of a curve defined by $x = f(t), y = g(t)$, $t \in [\alpha, \beta]$ with f', g' continuous and the curve traversed only once as t increases from α to β is

$$\int_{\alpha}^{\beta} \sqrt{\left(\frac{dx}{dt}\right)^2 + \left(\frac{dy}{dt}\right)^2} \, dt.$$

Since $(ds)^2 = (dx)^2 + (dy)^2$ (see section 9.1) this is $\int ds$.

5) Write a definite integral for the arc length of each:

a) the curve given by $x = \ln t, y = t^2$, for $1 \le t \le 2$.

$x'(t) = \frac{1}{t}$ and $y'(t) = 2t$. The arc length is

$$\int_{1}^{2} \sqrt{\left(\frac{1}{t}\right)^2 + (2t)^2} \, dt.$$

b) an ellipse given by $\dfrac{x^2}{a^2} + \dfrac{y^2}{b^2} = 1$, where $a, b > 0$.

The ellipse may be defined parametrically by $x = a \cos t, y = b \sin t, t \in [0, 2\pi]$. The length is

$$\int_{0}^{2\pi} \sqrt{(-a \sin t)^2 + (b \cos t)^2} \, dt$$

$$= \int_{0}^{2\pi} \sqrt{a^2 \sin^2 t + b^2 \cos^2 t} \, dt$$

c) The curve given by

$x = \cos 2t$

$y = \sin t$

$t \in [0, 2\pi]$

t	x	y
0	1	0
$\frac{\pi}{4}$	0	$\frac{\sqrt{2}}{2}$
$\frac{\pi}{2}$	-1	1
$\frac{3\pi}{4}$	0	$\frac{\sqrt{2}}{2}$
π	1	0
$\frac{5\pi}{4}$	0	$-\frac{\sqrt{2}}{2}$
$\frac{3\pi}{2}$	-1	-1
$\frac{7\pi}{4}$	0	$-\frac{\sqrt{2}}{2}$
2π	1	0

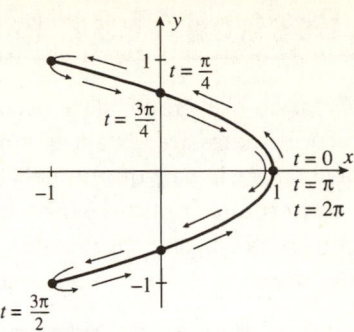

The curve is traversed once for $t \in \left[\frac{\pi}{2}, \frac{3\pi}{2}\right]$ and has length

$$\int_{\frac{\pi}{2}}^{\frac{3\pi}{2}} \sqrt{4 \sin^2 2t + \cos^2 t}\ dt.$$

Page 701
(ET Page 665)

D. If a curve given by $x = f(t)$, $y = g(t)$, $t \in [\alpha, \beta]$, with f', g' continuous and $g(t) \geq 0$ is rotated about the x-axis, the area of the resulting surface of revolution is

$$S = \int_{\alpha}^{\beta} 2\pi y \sqrt{\left(\frac{dx}{dt}\right)^2 + \left(\frac{dy}{dt}\right)^2}\ dt.$$

Using the ds notation this is $S = \int 2\pi y\ ds$, the same formula as in section 8.3.

6) Find the definite integral for the area of the surface of revolution about the x-axis for each curve:

a) $x = 2t + 1$, $y = t^3$, $t \in [0, 2]$

$$S = \int_0^2 2\pi(t^3)\sqrt{(2)^2 + (3t^2)^2}\ dt$$
$$= \int_0^2 2\pi t^3 \sqrt{4 + 9t^4}\ dt.$$

b) a "football" obtained from rotating the ellipse $\frac{x^2}{a^2} + \frac{y^2}{b^2} = 1$, $y \geq 0$ about the x-axis.

The top half of the ellipse is $x = a \cos t$, $y = b \sin t$, $t \in [0, \pi]$.
$$ds = \sqrt{(dx)^2 + (dy)^2}$$
$$= \sqrt{(-a \sin t)^2 + (b \cos t)^2}$$
$$= \sqrt{(a^2 \sin^2 t + b^2 \cos^2 t}.$$

Thus the area
$$S = \int_0^{\pi} 2\pi(b \sin t)\sqrt{a^2 \sin^2 t + b^2 \cos^2 t}\ dt.$$

Section 11.3 Polar Coordinates

Thus far all our graphs have been in a rectangular coordinate system where the two coordinates are distances from perpendicular axes. In this section on polar coordinates the pair of numbers that determines a point are an angle through which to rotate from the x-axis and a distance from the origin. We will see how to convert from one coordinate system to another and see that certain curves are much easier to express in polar coordinates than rectangular coordinates.

Concepts to Master

A. Points in polar coordinates; conversion to and from rectangular to polar coordinates

B. Graphs of equations in polar coordinates

C. Tangents to polar curves

Summary and Focus Questions

A. To construct a *polar coordinate system* start with a point called the *pole* and a ray from the pole called the *polar axis*.

Page 705 (ET Page 669)

Pole

A point P has polar coordinates (r, θ) if

$|r|$ = the distance form the pole to P, and

θ = the measure of a directed angel with initial side the polar axis and terminal side the line through the pole and P.

To plot a point P with polar coordinates $P(r, \theta)$:

(1) if $r = 0$, P is the pole (for any value of θ).

(2) if $r > 0$, rotate the polar axis by the angle θ (counterclockwise for $\theta > 0$, clockwise for $\theta < 0$) and locate P on this ray at a distance r units from the pole.

(3) if $r < 0$, rotate the polar axis by the angle θ and then reflect about the pole. P is located on this reflected ray at a distance $-r$ units from the pole.

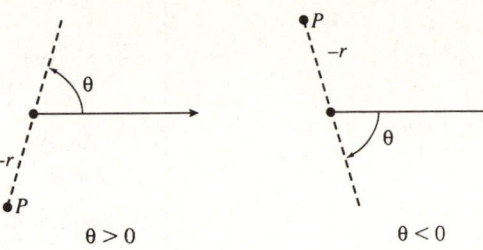

$\theta > 0$ $\theta < 0$

If a rectangular coordinate system is placed upon the polar coordinate system as in the figure, then to change from polar to rectangular:

$x = r \cos \theta$

$y = r \sin \theta$.

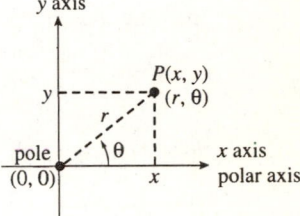

To change from rectangular to polar:

θ is a solution to $\tan \theta = \dfrac{y}{x}$ (for $x \neq 0$).

r is a solution to $r^2 = x^2 + y^2$ where $r > 0$ if the terminal side of the angle θ is in the same quadrant as P; if not, then $r \leq 0$.

1) Plot these points in polar coordinates.

A: $\left(4, \dfrac{\pi}{3}\right)$ B: $\left(-3, \dfrac{\pi}{4}\right)$

C: $(0, 3\pi)$ D: $\left(2, -\dfrac{\pi}{4}\right)$

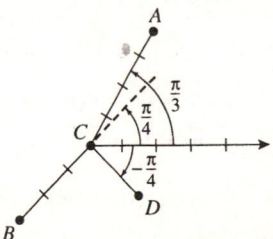

2) Find two other polar coordinates for point with polar coordinates $\left(8, \dfrac{\pi}{3}\right)$.

There are infinitely many answers including

$\left(8, \dfrac{7\pi}{3}\right), \left(-8, \dfrac{4\pi}{3}\right), \left(8, -\dfrac{5\pi}{3}\right), \ldots$

3) Sometimes, Always, or Never:

  a) The polar coordinates of a point are unique.

Never

  b) $(r, \theta) = (r, \theta + 2\pi)$.

Always

  c) $(r, \theta) = (-r, \theta + \pi)$.

Always

  d) $(r, \theta) = (-r, \theta)$.

Sometimes. (True when $r = 0$).

4) Find polar coordinates for the point P with rectangular coordinates $P: (-3, 3\sqrt{3})$.

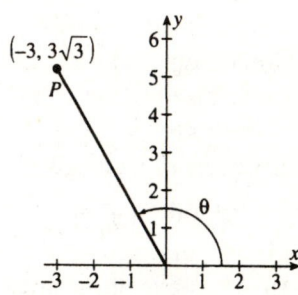

$\tan\theta = \dfrac{y}{x} = \dfrac{3\sqrt{3}}{-3} = -\sqrt{3}.$

A solution is $\theta = \dfrac{2\pi}{3}$.

$r^2 = (-3)^2 + (3\sqrt{3})^2 = 9 + 27 = 36.$

Because the terminal side of $\theta = \dfrac{2\pi}{3}$ lies in the same quadrant as P, $r > 0$. Therefore, $P: \left(6, \dfrac{2\pi}{3}\right)$. *Note:* $\theta = -\dfrac{\pi}{3}$ is another solution to $\tan\theta = -\sqrt{3}$, which results in $r = -6$ and coordinates $\left(-6, -\dfrac{\pi}{3}\right)$.

5) Find the rectangular coordinates for the point with polar coordinates $\left(-4, \dfrac{5\pi}{6}\right)$.

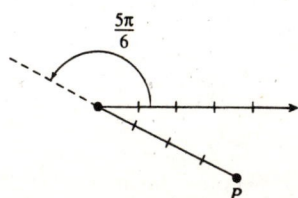

$x = -4 \cos\dfrac{5\pi}{6} = (-4)\left(\dfrac{-\sqrt{3}}{2}\right) = 2\sqrt{3}.$

$y = -4 \sin\dfrac{5\pi}{6} = (-4)\left(\dfrac{1}{2}\right) = -2.$

Thus $P: (2\sqrt{3}, -2).$

B. The procedure for graphing a polar equation is often the same as that you used when you first graphed functions—compute values and plot points. In some cases the graph of a polar equation is easily identified when the equation is transformed into rectangular coordinates. In other cases you may be able to take advantage of symmetry.

For example, $r = 6 \sin \theta$ becomes $r^2 = 6r \sin \theta$ or $x^2 + y^2 = 6y$. This is $x^2 + y^2 - 6y = 0$ or $x^2 + (y - 3)^2 = 9$, the circle of radius 3 centered at $(0, 3)$.

6) Sketch a graph of $r = \sin \theta - \cos \theta$.

Compute several points for values of θ.

A: $(-1, 0)$

B: $\left(\frac{1}{2} - \frac{\sqrt{3}}{2}, \frac{\pi}{6}\right)$

C: $\left(0, \frac{\pi}{4}\right)$

D: $\left(1, \frac{\pi}{2}\right)$

E: $\left(\sqrt{2}, \frac{3\pi}{4}\right)$

F: $(1, \pi)$ (same as A)

G: $\left(\frac{-1}{2} + \frac{\sqrt{3}}{2}, \frac{7\pi}{6}\right)$ (same as B)

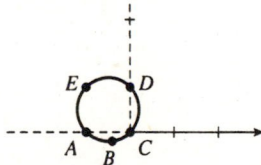

We can verify that the figure is a circle by switching to rectangular coordinates:

Multiply $r = \sin \theta - \cos \theta$ by r:

$r^2 = r \sin \theta - r \cos \theta$

$x^2 + y^2 = y - x$

$x^2 + x + y^2 - y = 0$

Completing the square gives

$\left(x + \frac{1}{2}\right)^2 + \left(y - \frac{1}{2}\right)^2 = \frac{1}{2},$

the circle with center $\left(-\frac{1}{2}, \frac{1}{2}\right)$ and radius $\frac{1}{\sqrt{2}}$.

7) Sketch the graph of $r = \cos 3\theta$

Since $\cos 3\theta = \cos(-3\theta)$, the curve is symmetric about the polar axis. Also $\cos 3\theta$ repeats every $\frac{2\pi}{3}$ units.

	θ	r
A	0	1
B	$\frac{\pi}{12}$	$\frac{\sqrt{2}}{2}$
C	$\frac{\pi}{6}$	0
D	$\frac{\pi}{4}$	$-\frac{\sqrt{2}}{2}$
E	$\frac{\pi}{3}$	-1
F	$\frac{5\pi}{12}$	$-\frac{\sqrt{2}}{2}$
G	$\frac{\pi}{2}$	0

	θ	r
H	$\frac{7\pi}{12}$	$\frac{\sqrt{2}}{2}$
I	$\frac{2\pi}{3}$	1
J	$\frac{3\pi}{4}$	$\frac{\sqrt{2}}{2}$
K	$\frac{5\pi}{6}$	0
L	$\frac{11\pi}{12}$	$-\frac{\sqrt{2}}{2}$
M	π	-1

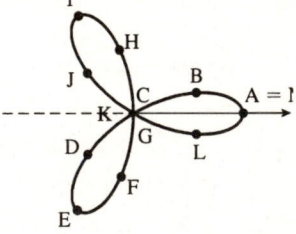

Page 710 (ET Page 674)

C. For polar curve $r = f(\theta)$, we can switch to rectangular coordinates $x = f(\theta)\cos\theta$, $y = f(\theta)\sin\theta$. Treating θ as a parameter,

$$\frac{dy}{dx} = \frac{\frac{dy}{d\theta}}{\frac{dx}{d\theta}} = \frac{\frac{dr}{d\theta}\sin\theta + r\cos\theta}{\frac{dr}{d\theta}\cos\theta - r\sin\theta}$$

This is the slope of the tangent line to $r = f(\theta)$ at (r, θ).

8) Find the slope of the tangent line to $r = e^\theta$ at $\theta = \frac{\pi}{2}$.

$x = r \cos \theta = e^\theta \cos \theta$, so

$\frac{dx}{d\theta} = e^\theta(-\sin \theta) + e^\theta(\cos \theta)$.

$y = r \sin \theta = e^\theta \sin \theta$, so

$\frac{dy}{d\theta} = e^\theta \cos \theta + e^\theta \sin \theta$.

At $\theta = \frac{\pi}{2}$,

$\frac{dx}{d\theta} = -e^{\pi/2}$ and $\frac{dy}{d\theta} = e^{\pi/2}$.

Therefore, $\frac{dy}{dx} = \frac{\frac{dy}{d\theta}}{\frac{dx}{d\theta}} = \frac{e^{\pi/2}}{-e^{\pi/2}} = -1$.

9) Find the points on $r = \sin \theta - \cos \theta$ where the tangent line is horizontal or vertical.

$\frac{dy}{dx} = \frac{\frac{dy}{d\theta}}{\frac{dx}{d\theta}}$ is not defined when $\frac{dx}{d\theta} = 0$

and zero when $\frac{dy}{d\theta} = 0$.

$x = (\sin \theta - \cos \theta) \cos \theta$

$\frac{dx}{d\theta} = (\sin \theta - \cos \theta)(-\sin \theta)$

$\qquad\qquad + \cos \theta \, (\cos \theta + \sin \theta)$

$\qquad = 2 \sin \theta \cos \theta + (\cos^2 \theta - \sin^2 \theta)$

$\qquad = \sin 2\theta + \cos 2\theta$.

$\frac{dx}{d\theta} = 0$ when

$\sin 2\theta = -\cos 2\theta$

$\tan 2\theta = -1$

$2\theta = \frac{3\pi}{4}$ and $\frac{7\pi}{4}$

$\theta = \frac{3\pi}{8}$ and $\frac{7\pi}{8}$.

At $\theta = \frac{3\pi}{8}$, $r = \sin \frac{3\pi}{8} - \cos \frac{3\pi}{8}$

$= \frac{\sqrt{\sqrt{2} + 2}}{2} - \frac{\sqrt{2 - \sqrt{2}}}{2} \approx .54$.

At $\theta = \frac{7\pi}{8}$, $r = \sin \frac{7\pi}{8} - \cos \frac{7\pi}{8}$

$= \frac{\sqrt{2 - \sqrt{2}}}{2} - \frac{-\sqrt{\sqrt{2} + 2}}{2} \approx 1.31$.

There are vertical tangents at $\left(.54, \frac{3\pi}{8}\right)$ and $\left(1.31, \frac{7\pi}{8}\right)$.

$$y = (\sin \theta - \cos \theta) \sin \theta$$

$$\frac{dy}{d\theta} = (\sin \theta - \cos \theta) \cos \theta$$
$$+ \sin \theta (\cos \theta + \sin \theta)$$
$$= 2 \sin \theta \cos \theta - (\cos^2 \theta - \sin^2 \theta)$$
$$= \sin 2\theta - \cos 2\theta = 0.$$

$$\sin 2\theta = \cos 2\theta$$

$$\tan 2\theta = 1$$

$$2\theta = \frac{\pi}{4} \text{ and } \frac{5\pi}{4}$$

$$\theta = \frac{\pi}{8} \text{ and } \frac{5\pi}{8}.$$

At $\theta = \frac{\pi}{8}, r = \sin \frac{\pi}{8} - \cos \frac{\pi}{8}$

$$= \frac{\sqrt{2 - \sqrt{2}}}{2} - \frac{\sqrt{2 + \sqrt{2}}}{2} \approx -.54.$$

At $\theta = \frac{5\pi}{8}, r = \sin \frac{5\pi}{8} - \cos \frac{5\pi}{8}$

$$= \frac{\sqrt{2 + \sqrt{2}}}{2} - \frac{-\sqrt{2 - \sqrt{2}}}{2} \approx 1.31.$$

There are horizontal tangents at $\left(-.54, \frac{\pi}{8}\right)$ and $\left(1.31, \frac{5\pi}{8}\right)$.

The graph is

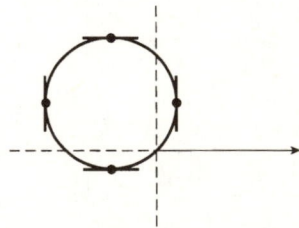

The graph is a circle.

$$r = \sin \theta - \cos \theta$$
$$r^2 = r \sin \theta - r \cos \theta$$
$$x^2 + y^2 = -x + y$$
$$x^2 + x + \frac{1}{4} + y^2 - y + \frac{1}{4} = \frac{1}{2}$$
$$\left(x + \frac{1}{2}\right)^2 + \left(y - \frac{1}{2}\right)^2 = \left(\frac{1}{\sqrt{2}}\right)^2.$$

In rectangular coordinates the center is $\left(-\frac{1}{2}, \frac{1}{2}\right)$ and the radius is $\frac{1}{\sqrt{2}}$.

Section 11.4 Areas and Lengths in Polar Coordinates

This section develops a formula for the area of a region bounded by equations given in polar form and the arc length of a curve given by a polar equation.

Concepts to Master

A. Area of a region described by polar equations
B. Length of a curve described by a polar equation

Summary and Focus Questions

Page 716 (ET Page 680)

A. The area of a region bounded by $\theta = a, \theta = b, r = f(\theta)$ where f is continuous and positive and $0 \le b - a \le 2\pi$ is $\int_a^b \frac{1}{2} [f(\theta)]^2 \, d\theta.$

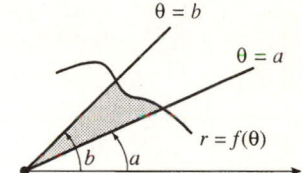

1) Find a definite integral for each region:

a) the shaded area

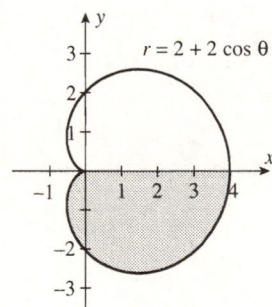

$r = 2 + 2 \cos \theta$

The region is bound by $\theta = \pi, \theta = 2\pi,$ $r = 2 + 2 \cos \theta.$ The area is

$$\int_\pi^{2\pi} \frac{1}{2}(2 + 2 \cos \theta)^2 \, d\theta$$

$$= 2 \int_\pi^{2\pi} (1 + \cos \theta)^2 \, d\theta.$$

b) the region bounded by $\theta = \frac{\pi}{3}, \theta = \frac{\pi}{2}$, $r = e^{\theta}$.

The area is $\int_{\pi/3}^{\pi/2} \frac{1}{2} e^{2\theta} \, d\theta$.

c) the shaded area.

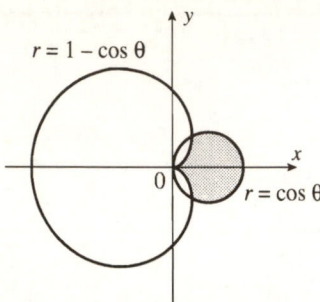

The area is between 2 curves so we must first determine where the curves intersect:

$1 - \cos\theta = \cos\theta, 1 = 2\cos\theta, \cos\theta = \frac{1}{2}$.

Therefore $\theta = \frac{\pi}{3}, -\frac{\pi}{3}$.

For $-\frac{\pi}{3} \le \theta \le \frac{\pi}{3}$,

$\cos\theta \ge 1 - \cos\theta$ so the area is

$$= \int_{-\frac{\pi}{3}}^{\frac{\pi}{3}} \frac{1}{2}[\cos\theta]^2 \, d\theta$$

$$- \int_{-\frac{\pi}{3}}^{\frac{\pi}{3}} \frac{1}{2}(1 - \cos\theta)^2 \, d\theta$$

$$= \int_{-\frac{\pi}{3}}^{\frac{\pi}{3}} \left(-\frac{1}{2} + \cos\theta\right) d\theta.$$

2) Find the area above the line $r = \csc\theta$ and inside the circle $r = 2$.

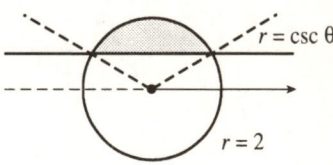

The curves intersect when $\csc\theta = 2$.

$\frac{1}{\sin\theta} = 2; \sin\theta = \frac{1}{2}. \theta = \frac{\pi}{6}$ and $\frac{5\pi}{6}$.

The area is $\int_{\frac{\pi}{6}}^{\frac{5\pi}{6}} \left(\frac{1}{2}(2)^2 - \frac{1}{2}\csc^2\theta\right) d\theta$

$$= \frac{1}{2} \int_{\frac{\pi}{6}}^{\frac{5\pi}{6}} (4 - \csc^2\theta) \, d\theta$$

$$= \frac{1}{2}(4\theta + \cot\theta)\Big|_{\frac{\pi}{6}}^{\frac{5\pi}{6}} = \frac{4\pi}{3} - \sqrt{3}.$$

B. A curve given in polar coordinates by $r = f(\theta)$ for $a \le \theta \le b$ has arc length

$$\int_a^b \sqrt{[f(\theta)]^2 + [f'(\theta)]^2}\, d\theta.$$

2) Set up a definite integral for the length of each curve.

a) the curve $r = e^{2\theta}$ for $0 \le \theta \le 1$.

$$\int_0^1 \sqrt{(e^{2\theta})^2 + (2e^{2\theta})^2}\, d\theta = \sqrt{5}\int_0^1 e^{2\theta}\, d\theta.$$

b) the curve sketched below.

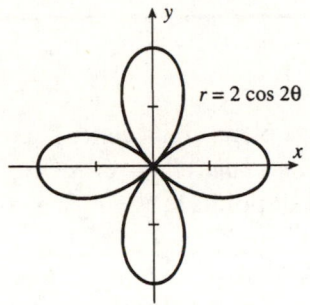

$r = 2\cos 2\theta$

Since the curve is quite symmetric, the arc length is 8 times the length of half of one leaf:

$$8\int_0^{\frac{\pi}{4}} \sqrt{[2\cos 2\theta]^2 + [-4\sin 2\theta]^2}\, d\theta$$

$$= 8\int_0^{\frac{\pi}{4}} \sqrt{4\cos^2 2\theta + 16\sin^2 2\theta}\, d\theta$$

$$= 16\int_0^{\frac{\pi}{4}} \sqrt{1 + 3\sin^2 2\theta}\, d\theta$$

Section 11.5 Conic Sections

Conic sections are the various curves resulting from the intersection of a plane with a cone. This section gives geometric definitions of the conic sections and their equations.

Concepts to Master

Focus-directrix definition of parabolas, ellipses and hyperbolas; vertices; standard form of the equations of conics

Summary and Focus Questions

Page 720 (ET Page 684)

Conic sections are curves obtained by intersecting a plane and a cone. There are several possibilities; the resulting curve is either a *parabola*, *ellipse* or *hyperbola*. These curves may also be obtained as a certain set of points satisfying a given geometric condition:

Parabola: Given a line (a *directrix*) and a point (the *focus*, F), a point P is on the parabola if the distance from P to the directrix is the same as the distance from P to F.

Ellipse: Given two points (the *foci*, F_1 and F_2), a point P is on the ellipse if the sum of the distances from P to F_1 and from P to F_2 is a constant.

Hyperbola: Given two points (the *foci*, F_1 and F_2), a point P is on the hyperbola if the difference of the distances from P to F_1 and from P to F_2 is a constant.

The equations of the conics and their graphs in rectangular coordinates are given below.

Conic Section	Equation	Properties	Graphs
Parabola	$x^2 = 4py$	focus: $(0, p)$ vertex: $(0, 0)$ directrix: $y = -p$	
		focus: $(0, p)$ vertex: $(0, 0)$ directrix: $y = -p$	

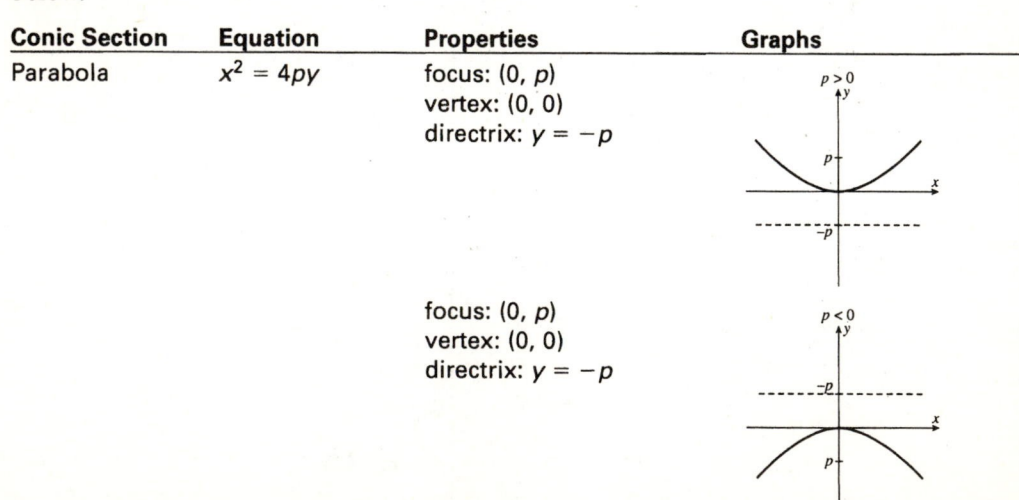

Parabola	$y^2 = 4px$	focus: $(p, 0)$ vertex: $(0, 0)$ directrix: $x = -p$	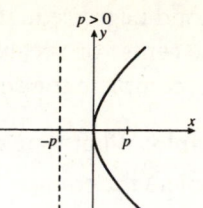
		focus: $(p, 0)$ vertex: $(0, 0)$ directrix: $x = -p$	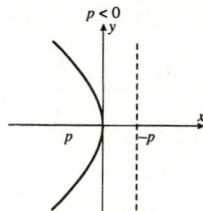
Ellipse	$\dfrac{x^2}{a^2} + \dfrac{y^2}{b^2} = 1$	foci: $(c, 0)$, $(-c, 0)$ vertices: $(a, 0)$, $(-a, 0)$ constant sum $= 2a$ center: $(0, 0)$ $c^2 = a^2 - b^2$, $a \geq b > 0$	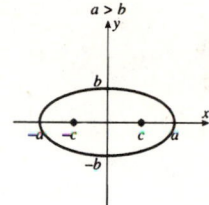
Ellipse	$\dfrac{x^2}{b^2} + \dfrac{y^2}{a^2} = 1$	foci: $(0, c)$, $(0, -c)$ vertices: $(0, a)$, $(0, -a)$ constant sum $= 2a$ center: $(0, 0)$ $c^2 = a^2 - b^2$, $a \geq b > 0$	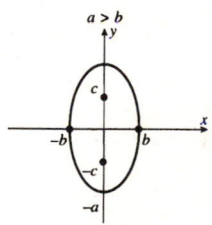
Hyperbola	$\dfrac{x^2}{a^2} - \dfrac{y^2}{b^2} = 1$	foci: $(c, 0)$, $(-c, 0)$ vertices: $(a, 0)$, $(-a, 0)$ constant difference $= 2a$ center $(0, 0)$ $c^2 = a^2 + b^2$ asymptotes: $y = \pm\dfrac{b}{a}x$	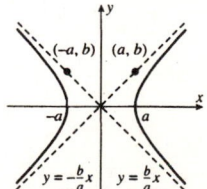
Hyperbola	$\dfrac{y^2}{a^2} - \dfrac{x^2}{b^2} = 1$	foci: $(0, c)$, $(0, -c)$ vertices: $(0, a)$, $(0, -a)$ constant difference $= 2a$ center: $(0, 0)$ $c^2 = a^2 + b^2$ asymptotes: $y = \pm\dfrac{a}{b}x$	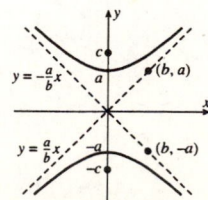

A second degree equation in x and y (with no xy term)represents a conic section whose center or vertex may be shifted from $(0, 0)$. To determine the type of conic, complete the square for x and y as in this example.

Example: What type of conic is given by the equation $x^2 + 6x + 4y^2 - 8y = 3$?
Complete the square:
$$x^2 + 6x + 9 + 4(y^2 - 2y + 1) = 3 + 9 + 4(1).$$
$$(x + 3)^2 + 4(y - 1)^2 = 16$$
$$\frac{(x + 3)^2}{16} + \frac{(y - 1)^2}{4} = 1.$$

This is an ellipse shifted 3 units left and one unit upward. Its center is $(-3, 1)$, $a = 4$ and $b = 2$.

1) Find the vertices, foci, and directrix (if a parabola) and sketch the graph of each:

a) $\frac{x^2}{144} = 1 + \frac{y^2}{25}$

$$\frac{x^2}{144} - \frac{y^2}{25} = 1.$$

This is a hyperbola with $a = 12$, $b = 5$.
$c^2 = 12^2 + 5^2 = 169$, $c = 13$.
Foci: $(13, 0)$, $(-13, 0)$
Vertices: $(12, 0)$, $(-12, 0)$
Asymptotes: $y = \frac{5}{12}x$, $y = -\frac{5}{12}x$.

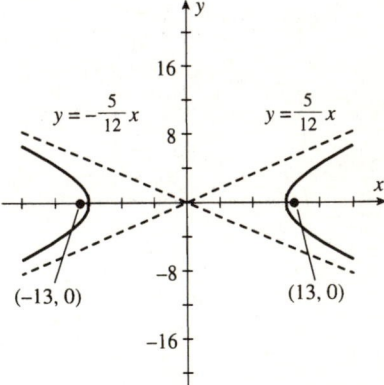

b) $x^2 = 4x + 8y - 4$

$x^2 = 4x + 8y - 4$
$x^2 - 4x + 4 = 8y$
$(x - 2)^2 = 4(2)y$
This is a parabola, shifted two units to the right. The vertex is $(2, 0)$.
$p = 2$; the directrix is $y = -2$.
The focus is $(2, 2)$, also shifted two units.

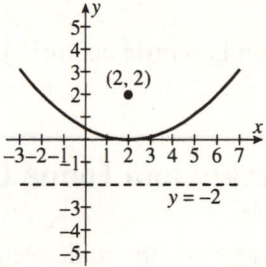

2) What is the conic given by
$2x(6 - x) = y(8 + y)$?

$2x(6 - x) = y(8 + y)$
$2x^2 + 12x = y^2 + 8y$
$2x^2 - 12x + y^2 + 8y = 0$
$2(x^2 - 6x + 9) + y^2 + 8y + 16 = 18 + 16$
$2(x - 3)^2 + (y + 4)^2 = 34$

$$\frac{(x - 3)^2}{17} + \frac{(y + 4)^2}{34} = 1$$

This is an ellipse , $a = \sqrt{34}, b = \sqrt{17}$.

Section 11.6 Conic Sections in Polar Coordinates

This section gives alternate definitions for each of the three conic sections that use just one focus and one directrix. This approach leads to a simple form for the equations of parabolas, ellipses and hyperbolas in polar coordinates.

Concepts to Master

Eccentricity; Eccentric definition of conic sections; Equations of conic in polar form

Summary and Focus Questions

Page 728 (ET Page 692)

Another way to define conic sections is to specify a fixed point F (the focus), a fixed line l (the directrix), and a positive constant e called the *eccentricity**. Then the set of all points P such that

$$\frac{\text{distance from } P \text{ to } F}{\text{distance from } P \text{ to } l} = e$$

is a conic section. The value of e determines whether the conic is a parabola, ellipse or hyperbola:

Eccentricity	Type	Graph	
$e = 1$	Parabola		$\frac{|PF|}{|Pl|} = 1$
$e < 1$	Ellipse		$\frac{|PF|}{|Pl|} = e < 1$
$e > 1$	Hyperbola		$\frac{|PF|}{|Pl|} = e > 1$

* This use of e for eccentricity is not to be confused with the use of e as the symbol for the base of natural logarithms.

Suppose a conic with eccentricity $e > 0$ is drawn in a rectangular coordinate system with focus F at $(0, 0)$, and one of the lines $x = d$, $x = -d$, $y = d$, or $y = -d$, where $d > 0$, is the directrix l. By superimposing a polar coordinate system the polar equation of the conic has one of these forms:

Conic	**Polar Form of Equation**			
Equation:	$r = \dfrac{ed}{1 + e \cos \theta}$	$r = \dfrac{ed}{1 - e \cos \theta}$	$r = \dfrac{ed}{1 + e \sin \theta}$	$r = \dfrac{ed}{1 - e \sin \theta}$
Directrix:	$x = d$	$x = -d$	$y = d$	$y = -d$
Parabola ($e = 1$)	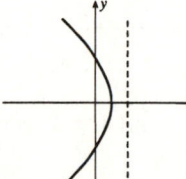	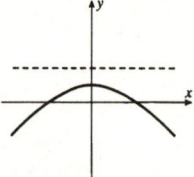		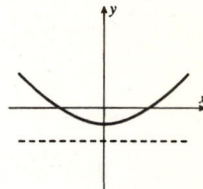
Ellipse ($0 < e < 1$)	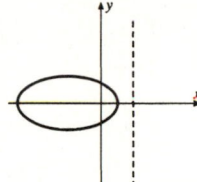	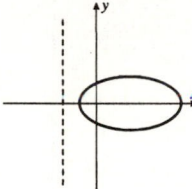	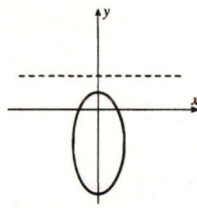	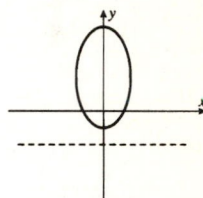
Hyperbola ($e > 1$)	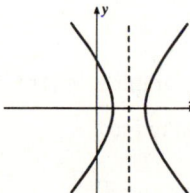	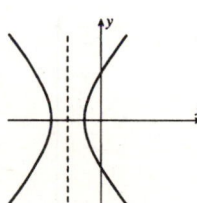	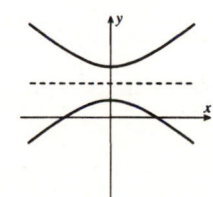	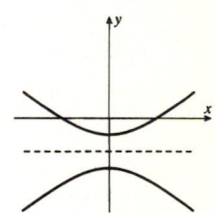

1) The sketch below shows one point P on a conic and its distances from the focus and directrix. What type of conic is it?

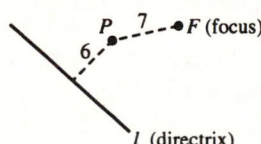

$e = \dfrac{|PF|}{|Pl|} = \dfrac{7}{6} > 1$, so the conic is a hyperbola.

2) What is the polar equation of the conic with:

a) eccentricity 2, directrix $y = 3$.

$e = 2$ and $d = 3$, so $r = \dfrac{6}{1 + 2 \sin \theta}$.

b) eccentricity $\frac{1}{2}$, directrix $x = -3$.

$e = \frac{1}{2}$ and $d = 3$, so $r = \dfrac{\frac{3}{2}}{1 - \frac{1}{2} \cos \theta}$,

$r = \dfrac{3}{2 - \cos \theta}$.

c) directrix $x = 4$ and is a parabola.

$e = 1$ and $d = 4$, so $r = \dfrac{4}{1 + \cos \theta}$.

3) What polar form does the equation of the graph below have?

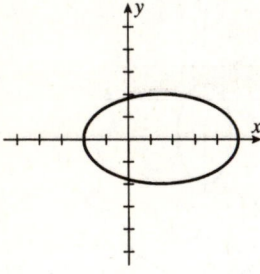

The graph is that of an ellipse. The directrix has the form $x = -d$ so the equation is

$$r = \dfrac{ed}{1 - e \cos \theta}.$$

4) Find the eccentricity and directrix and identify the conic given by $r = \dfrac{3}{4 - 5 \cos \theta}$.

Divide numerator and denominator by 4.

$r = \dfrac{\frac{3}{4}}{1 - \frac{5}{4} \cos \theta}$, $e = \frac{5}{4}$. Since $ed = \frac{3}{4}$,

$d = \dfrac{3}{4} \cdot \dfrac{1}{e} = \dfrac{3}{4} \cdot \dfrac{4}{5} = \dfrac{3}{5}$.

The trigonometric term is $-\cos \theta$ so the directrix is $x = -\frac{3}{5}$. Since $e > 1$, the conic is a hyperbola.

Technology Plus for Chapter 11

1) Use a graphing calculator to sketch a graph
of $x = t + 2 \sin 2t$, $y = t + \sin 4t$
for $-9 \le t \le 9$.

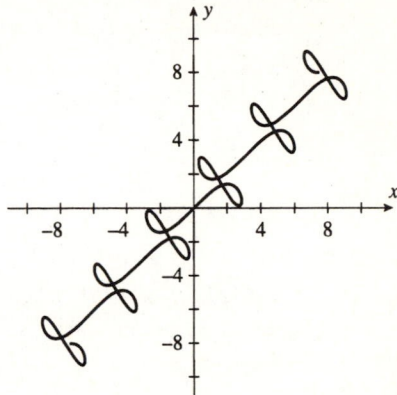

2) Use a CAS to evaluate the definite integral
for the arc length of
$x = t + \cos t$
$y = t + \sin t$
for $0 \le t \le 2\pi$.

$\dfrac{dx}{dt} = 1 - \sin t$, $\dfrac{dy}{dt} = 1 + \cos t$

The arc length is
$$\int_0^{2\pi} \sqrt{(1 - \sin t)^2 + (1 + \cos t)^2}\, dt$$
≈ 10.037.

3) Sketch the graphs of $r = \cos (2k - 1)\theta$,

$0 \le \theta \le 2\pi$, for $k = 1, 2, 3$ and 4.

How does the graph change as k increases?

$k = 1$

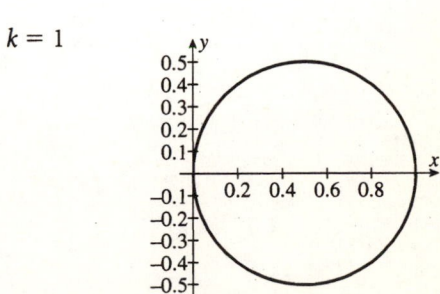

$k = 2$

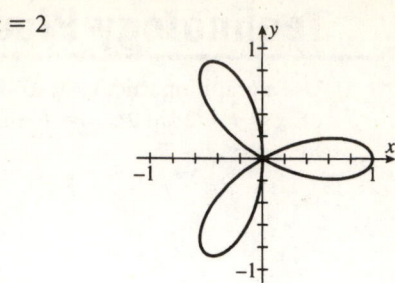

$k = 3$

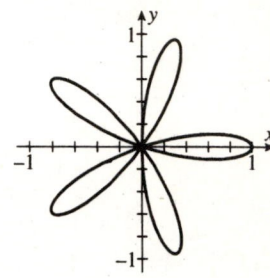

$k = 4$

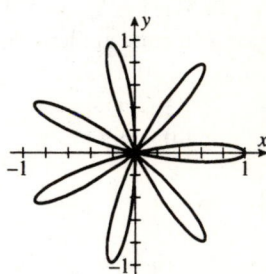

The number of leaves is $2k + 1$.

4) Sketch a graph of $r = \dfrac{1}{1 - \frac{1}{2}\cos\theta}$ on a graphing calculator. What type of conic is it?

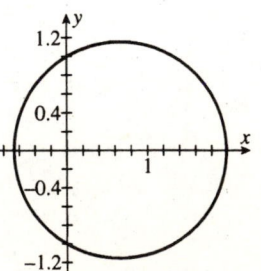

The graph is an ellipse $(e = \frac{1}{2})$.

Chapter 12 — Infinite Sequences and Series

"I'M BEGINNING TO UNDERSTAND ETERNITY, BUT INFINITY IS STILL BEYOND ME."

Cartoons courtesy of Sidney Harris. Used by permission.

Section 12.1   Sequences

The main topic of this chapter is the representation of function as a series. A series (introduced in section 11.2) is a sum of an infinite list of numbers—that is, the sum of a sequence of numbers. This first section describes the basic concepts for sequences.

Concepts to Master

A. Sequences; Limit of a sequence, convergence and divergence

B. Monotone sequence (increasing, decreasing); Bounded sequence; Monotonic Sequence Theorem

C. Sequences defined recursively

Summary and Focus Questions

Page 737 (ET Page 701)*

A. A *sequence* is an infinite list of numbers given in a specific order:

$$\{a_n\} = a_1, a_2, a_3, \ldots, a_n, a_{n+1}, \ldots.$$

a_n is called the nth *term* of the sequence.

The sequence $\{a_n\}$ *converges* to a number L, written $\lim\limits_{n\to\infty} a_n = L$, means that the values of a_n get closer and closer to L as n grows larger. The formal definition is:

$\lim\limits_{n\to\infty} a_n = L$ means for all $\epsilon > 0$ there exists a positive integer N such that $|a_n - L| < \epsilon$ for all $n > N$.

If $\lim\limits_{n\to\infty} a_n$ does not exist, we say $\{a_n\}$ *diverges*.

$\lim\limits_{n\to\infty} a_n = \infty$ means the a_n terms grow without bound.

One way to evaluate $\lim\limits_{n\to\infty} a_n$ is to find a real function $f(x)$ such that $f(n) = a_n$ for all n. If $\lim\limits_{n\to\infty} f(x) = L$, then $\lim\limits_{n\to\infty} a_n = L$. The converse is false.

All limit laws for limits of functions at infinity are valid for convergent sequences. A version of the Squeeze Theorem also holds:

If $a_n \le b_n \le c_n$ for all $n \ge N$ and $\lim\limits_{n\to\infty} a_n = \lim\limits_{n\to\infty} c_n = L$, then $\lim\limits_{n\to\infty} b_n = L$.

1) Find the fourth term of the sequence
$$a_n = \frac{(-1)^n}{n^2}.$$

$$a_4 = \frac{(-1)^4}{4^2} = \frac{1}{16}.$$

2) $\lim\limits_{n\to\infty} x_n = K$ means
for all ____ there
exists ____ such
that ____ for all
____.

$\epsilon > 0$
positive integer N
$|x_n - K| < \epsilon$
$n > N.$

3) Determine whether each converges.

   a) $a_n = \dfrac{n}{n^2 + 1}.$

$$\lim\limits_{n\to\infty} \frac{n}{n^2 + 1} = \text{(divide by } n^2)$$

$$\lim\limits_{n\to\infty} \frac{\frac{1}{n}}{1 + \frac{1}{n^2}} = \frac{0}{1 + 0} = 0.$$

Thus a_n converges to 0.

   b) $b_n = \dfrac{n^2 + 1}{2n}.$

b_n grows without bound, so $\{b_n\}$ diverges. In this case we may write
$$\lim\limits_{n\to\infty} \frac{n^2 + 1}{2n} = \infty.$$

   c) $c_n = \dfrac{(-1)^n n}{n + 1}.$

The sequence $\dfrac{n}{n+1}$ converges to 1 so $c_n = \dfrac{(-1)^n n}{n + 1}$ alternately takes on values near 1, and -1. Thus $\{c_n\}$ diverges.

   d) $d_n = \dfrac{n}{n + 1} + \dfrac{2}{n^2}.$

$$\lim\limits_{n\to\infty} d_n = \lim\limits_{n\to\infty} \frac{n}{n + 1} + \lim\limits_{n\to\infty} \frac{2}{n^2} = 1 + 0 = 1.$$

4) Find $\lim\limits_{n\to\infty} \dfrac{\cos n}{n}.$

Since $-1 \le \cos n \le 1$ for all n,
$$-\frac{1}{n} \le \frac{\cos n}{n} \le \frac{1}{n} \text{ for all } n.$$

Because both $\left\{-\frac{1}{n}\right\}$ and $\left\{\frac{1}{n}\right\}$ converge to 0,
$$\lim\limits_{n\to\infty} \frac{\cos n}{n} = 0.$$

5) Find $\lim\limits_{n\to\infty} \dfrac{\sin x}{n}$.

Since $\sin x$ is a constant as far as n is concerned, $\lim\limits_{n\to\infty} \dfrac{\sin x}{n} = (\sin x) \lim\limits_{n\to\infty} \dfrac{1}{n}$
$= (\sin x)(0) = 0.$

Silliness: Don't conclude the limit is 6 with this "computation":

$$\lim_{n\to\infty} \frac{\sin x}{n} = \lim_{n\to\infty} \frac{\sin x}{\cancel{n}}$$
$$= \lim_{n\to\infty} \text{six} = 6.$$

6) Find $\lim\limits_{n\to\infty} \dfrac{\ln n}{n}$.

Let $f(x) = \dfrac{\ln x}{x}$.

$\lim\limits_{x\to\infty} \dfrac{\ln x}{x}$ $\left(\dfrac{\infty}{\infty} \text{ form, L'Hospital's Rule}\right)$

$= \lim\limits_{x\to\infty} \dfrac{\frac{1}{x}}{1} = 0.$

Thus $\lim\limits_{n\to\infty} \dfrac{\ln n}{n} = 0.$

7) True, False:

   a) If $\{a_n\}$ and $\{b_n\}$ converge, then $\{a_n + b_n\}$ converges.

True.

   b) If $\{a_n\}$ and $\{b_n\}$ diverge, then $\{a_n + b_n\}$ diverges.

False. For example, $a_n = n^2$ and $b_n = -n^2$ diverge, but $\{a_n + b_n\}$ is the constant sequence of all zeros that converges to zero.

8) If $\lim\limits_{n\to\infty} s_n = 4$ and $\lim\limits_{n\to\infty} t_n = 2$, then

   a) $\lim\limits_{n\to\infty} (8s_n - 2t_n) = \underline{\hspace{1cm}}.$

$8(4) - 2(2) = 28.$

   b) $\lim\limits_{n\to\infty} \dfrac{3s_n}{t_n} = \underline{\hspace{1.5cm}}.$

$\dfrac{3(4)}{2} = 6.$

Page
744
(ET Page
708)

B. $\{a_n\}$ is *increasing* means $a_{n+1} \geq a_n$ for all n.
$\{a_n\}$ is *decreasing* means $a_{n+1} \leq a_n$ for all n.
$\{a_n\}$ is *monotonic* if it is either increasing or decreasing.
One way to show that $\{a_n\}$ is increasing is to show that $\dfrac{da_n}{dn} \geq 0$ (treating n as a real number and a_n as a function of n).
$\{a_n\}$ is *bounded above* means $a_n \leq M$ for some M and all n.
$\{a_n\}$ is *bounded below* means $a_n \geq m$ for some m and all n.
$\{a_n\}$ is *bounded* if it is both bounded above and bounded below.

Monotonic Sequence Theorem: If $\{a_n\}$ is bounded and monotonic, then $\{a_n\}$ converges.

For example, $a_n = \dfrac{1}{e^n + 1}$ is monotonic (decreasing) and bounded ($a_n > 0$ for all n). Therefore $\lim\limits_{n \to \infty} \dfrac{1}{e^n + 1}$ exists (It is 0.)

The converse of the Monotonic Sequence Theorem is "partially true":

>   If $\{a_n\}$ converges, then $\{a_n\}$ is bounded.

>   However, a convergent sequence need not be monotonic.

9) Is $c_n = \dfrac{1}{3n}$ bounded above? bounded below?

$\left\{\dfrac{1}{3n}\right\}$ is bounded above by $\dfrac{1}{3}$ and bounded below by 0.

10) Is $s_n = \dfrac{n}{n+1}$ an increasing sequence?

Yes, because
$$\frac{ds_n}{dn} = \frac{(n+1) - n}{(n+1)^2} = \frac{1}{(n+1)^2} > 0.$$

11) True or False:

a) If $\{a_n\}$ is not bounded below then $\{a_n\}$ diverges.

True.

b) If $\{a_n\}$ is decreasing and $a_n \geq 0$ for all n then $\lim\limits_{n \to \infty} a_n$ exists.

True.

c) If $\{a_n\}$ is bounded, then $\{a_n\}$ converges.

False. For example, $a_n = (-1)^n$.

Page
745
(ET Page
709)

C. A sequence $\{a_n\}$ is *defined by a recurrence relation* (defined *recursively*) means:

1. a_1 is defined.

2. a_{n+1}, for $n = 1,2,3, \ldots$, is defined in terms of previous a_i. (Often a_{n+1} is defined using only a_n.)

For example, the sequence $\{a_n\}$ given by

$$a_1 = \frac{1}{2}$$
$$a_{n+1} = \frac{a_n}{2}, n = 1, 2, 3, \ldots$$

is the sequence $\frac{1}{2}, \frac{1}{4}, \frac{1}{8}, \frac{1}{16}, \ldots$. This is a recursive definition of $a_n = \frac{1}{2^n}$.

---

12) Define the sequence $\frac{1}{2}, \frac{3}{4}, \frac{7}{8}, \frac{15}{16}, \frac{31}{32}, \ldots$ with a recurrence relation.

$a_1 = \frac{1}{2}.$ $a_2 = \frac{3}{4} = \frac{1}{2} + \frac{1}{4} = a_1 + \frac{1}{4}.$

$a_3 = \frac{7}{8} = \frac{3}{4} + \frac{1}{8} = a_2 + \frac{1}{8}.$

The pattern shows that

$$a_{n+1} = a_n + \frac{1}{2^{n+1}}, \text{ for } n = 1, 2, 3, \ldots$$

(Note: a non-recursive answer is

$$a_n = \frac{2^n - 1}{2^n}.)$$

13) Let $a_1 = 2$ and $a_{n+1} = \frac{2a_n}{3}$. Find $\lim_{n \to \infty} a_n$.

$a_1 = 2, a_2 = \frac{2(2)}{3} = \frac{4}{3},$

$a_3 = \frac{2\left(\frac{4}{3}\right)}{3} = \frac{8}{9}, a_4 = \frac{2\left(\frac{8}{9}\right)}{3} = \frac{16}{27},$

$a_5 = \frac{2\left(\frac{16}{27}\right)}{3} = \frac{32}{81}.$

Since the denominator is growing faster than the numerator, it seems that $\lim_{n \to \infty} a_n = 0.$ This is so since a non-recursive formula for a_n is $2\left(\frac{2}{3}\right)^{n-1}.$

14) Suppose $\{a_n\}$ is defined as $a_1 = 2$, $a_2 = 1$ and $a_{n+2} = a_{n+1} - a_n$ for $n = 1, 2, 3, \ldots$. Find $\lim_{n \to \infty} a_n$.

Calculate a few terms to understand the pattern.

$a_1 = 2, a_2 = 1$
$a_3 = 1 - 2 = -1$
$a_4 = -1 - 1 = -2$
$a_5 = -2 - (-1) = -1$
$a_6 = -1 - (-2) = 1$
$a_7 = 1 - (-1) = 2$
$a_8 = 2 - 1 = 1$
$a_9 = 1 - 2 = -1$
$\vdots$

The terms cycle through $1, -1, \ldots$ and eventually return to $1, -1$. $\lim_{n \to \infty} a_n$ does not exist.

Section 12.2   Series

A series is the sum of all the terms of an infinite sequence. To determine whether (and if so, to what) a series sums, we build a second sequence of sums part way through the sequence, the first term, the sum of the first two terms, the sum of the first three terms, the sum of the first four terms, etc. The sum of the series exists (that is, the series converges) if the limit of this partial sum sequence exists. In this section we define these concepts precisely and look at some specific series which converge and others which do not.

Concepts to Master

A. Infinite series, Partial sums; Convergent and divergent series; Convergent series laws

B. Geometric series; Value of a converging geometric series; Harmonic series

Summary and Focus Questions

Page
749
(ET Page
713)

A. Adding up all the terms of a sequence $\{a_n\}$ is an *(infinite) series*:

$$\sum_{k=1}^{\infty} a_k = a_1 + a_2 + a_3 + \ldots + a_n + \ldots$$

If we add up just the first n terms, we have the *nth partial sum* of the series:

$$s_n = \sum_{k=1}^{n} a_k = a_1 + a_2 + a_3 + \ldots + a_n$$

A series *converges* if the limit of its sequence of partial sums exists. In other words $\sum_{n=1}^{\infty} a_n$ *converges* (to s) means $\lim_{n \to \infty} s_n$ exists (and is s).

If $\lim_{n \to \infty} s_n$ does not exist, then $\sum_{n=1}^{\infty} a_n$ *diverges*.

Example: Let $a_n = \frac{1}{2^n}$. Then, $s_1 = a_1 = \frac{1}{2}$. $s_2 = a_1 + a_2 = \frac{1}{2} + \frac{1}{4} = \frac{3}{4}$.

$s_3 = a_1 + a_2 + a_3 = \frac{1}{2} + \frac{1}{4} + \frac{1}{8} = \frac{7}{8}$. s_n is $\frac{2^n - 1}{2^n}$, which converges to 1.

Thus, $\sum_{n=1}^{\infty} \frac{1}{2^n} = 1$.

Test for Divergence:

If $\lim_{n \to \infty} a_n$ does not exist or $\lim_{n \to \infty} a_n \neq 0$, $\sum_{n=1}^{\infty} a_n$ diverges.

Thus, for example, the series $\sum_{n=1}^{\infty} 2^n$ diverges, since $\lim_{n \to \infty} 2^n \neq 0$.

However, just because the terms a_n approach zero is not enough to conclude that $\sum_{n=1}^{\infty} a_n$ converges.

If $\sum\limits_{n=1}^{\infty} a_n$ converges (to L) and $\sum\limits_{n=1}^{\infty} b_n$ converges (to M) then:

$$\sum_{n=1}^{\infty} (a_n \pm b_n) \text{ converges (to } L \pm M).$$

$$\sum_{n=1}^{\infty} ca_n \text{ converges (to } cL) \text{ for } c \text{ any constant.}$$

1) A series converges if the _____ of the sequence of _____ exists.

limit, partial sums

2) Find the first four partial sums of $\sum\limits_{n=1}^{\infty} \dfrac{1}{n^2}$.

$$s_1 = \frac{1}{1^2} = 1.$$

$$s_2 = \frac{1}{1^2} + \frac{1}{2^2} = 1 + \frac{1}{4} = \frac{5}{4}.$$

$$s_3 = \frac{1}{1^2} + \frac{1}{2^2} + \frac{1}{3^2} = \frac{5}{4} + \frac{1}{9} = \frac{49}{36}.$$

$$s_4 = \frac{1}{1^2} + \frac{1}{2^2} + \frac{1}{3^2} + \frac{1}{4^2} = \frac{49}{36} + \frac{1}{16} = \frac{205}{144}.$$

3) Sometimes, Always, or Never:

   a) If $\lim\limits_{n\to\infty} a_n = 0$, $\sum\limits_{n=1}^{\infty} a_n$ converges.

Sometimes.

   b) If $\lim\limits_{n\to\infty} a_n \neq 0$, $\sum\limits_{n=1}^{\infty} a_n$ diverges.

Always.

4) Suppose $\sum\limits_{n=1}^{\infty} a_n = 3$ and $\sum\limits_{n=1}^{\infty} b_n = 4$.

Evaluate each of the following:

   a) $\sum\limits_{n=1}^{\infty} (a_n + 2b_n)$

$$3 + 2(4) = 11.$$

   b) $\sum\limits_{n=1}^{\infty} \dfrac{a_n}{5}$

$$\frac{3}{5}.$$

   c) $\sum\limits_{n=1}^{\infty} \dfrac{1}{a_n}$

This diverges. If $\sum\limits_{n=1}^{\infty} a_n = 3$, then $\lim\limits_{n\to\infty} a_n = 0$. Thus $\lim\limits_{n\to\infty} \dfrac{1}{a_n} \neq 0$, so $\sum\limits_{n=1}^{\infty} \dfrac{1}{a_n}$ diverges.

5) Sometimes, Always, or Never:

a) $\displaystyle\sum_{n=1}^{\infty} a_n$ converges, but $\displaystyle\lim_{n\to\infty} a_n$ does not exist.

Never. If $\displaystyle\sum_{n=1}^{\infty} a_n$ converges, then $\displaystyle\lim_{n\to\infty} a_n$ exists and is 0.

b) $\displaystyle\sum_{n=1}^{\infty} a_n$ converges, $\displaystyle\sum_{n=1}^{\infty} (a_n + b_n)$ converges, but $\displaystyle\sum_{n=1}^{\infty} b_n$ diverges.

Never. $\displaystyle\sum_{n=1}^{\infty} b_n = \sum_{n=1}^{\infty} (a_n + b_n) - \sum_{n=1}^{\infty} a_n$ is the difference of two convergent series and must converge.

6) Does $\displaystyle\sum_{n=1}^{\infty} \cos\left(\frac{1}{n}\right)$ converge?

No, the nth term does not approach 0.

Page 751
(ET Page 715)

B. A *geometric series* has the form

$$\sum_{n=1}^{\infty} ar^{n-1} = a + ar + ar^2 + ar^3 + \dots$$

and converges (for $a \neq 0$) to $\frac{a}{1-r}$ if and only if $-1 < r < 1$.

The *harmonic series* $\displaystyle\sum_{n=1}^{\infty} \frac{1}{n} = 1 + \frac{1}{2} + \frac{1}{3} + \dots + \frac{1}{n} + \dots$ diverges. This is an example of a series where $\displaystyle\lim_{n\to\infty} a_n = 0$ but $\displaystyle\sum_{n=1}^{\infty} a_n$ diverges.

7) a) $\displaystyle\sum_{n=1}^{\infty} \frac{2^n}{100}$ is a geometric series in which

$a = $ _____ and $r = $ _____.

$$\sum_{n=1}^{\infty} \frac{2^n}{100} = \frac{2}{100} + \frac{4}{100} + \frac{8}{100} + \dots$$

$a = \frac{2}{100}$ and $r = 2$.

b) Does the series converge?

No, since $r \geq 1$.

8) $\displaystyle\sum_{n=1}^{\infty} \left(\frac{-2}{9}\right)^n = $ _____.

This is a geometric series with $a = r = -\frac{2}{9}$ which converges to $\frac{a}{1-r} = -\frac{2}{11}$.

9) Does $\sum\limits_{n=1}^{\infty} \dfrac{6}{n}$ converge?

It diverges, since it is a multiple of the harmonic series:

$$\sum_{n=1}^{\infty} \frac{6}{n} = 6 \sum_{n=1}^{\infty} \frac{1}{n}.$$

10) $\sum\limits_{n=1}^{\infty} \dfrac{2^n}{3^{n+1}} = $ _____.

$\sum\limits_{n=1}^{\infty} \dfrac{2^n}{3^{n+1}} = \sum\limits_{n=1}^{\infty} \dfrac{2}{9}\left(\dfrac{2}{3}\right)^{n-1}$, which is a

geometric series with $a = \dfrac{2}{9}$, $r = \dfrac{2}{3}$ that

converges to $\dfrac{\frac{2}{9}}{1 - \frac{2}{3}} = \dfrac{2}{3}$.

11) Achilles gives the tortoise a 100 m head start. If Achilles runs at 5 m/s and the tortoise at $\frac{1}{2}$ m/s how far has the tortoise traveled by the time Achilles catches him?

Achilles

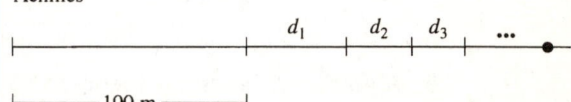

Tortoise

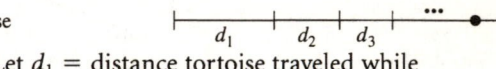

Let d_1 = distance tortoise traveled while Achilles was running the 100 m to the tortoise's starting point.

$d_1 = \left(\dfrac{100 \text{ m}}{5 \text{ m/s}}\right)\left(\dfrac{1}{2} \text{ m/s}\right) = 10$ m.

Let d_2 = distance tortoise traveled while Achilles was running the distance d_1. Since $d_1 = 10$ and Achilles runs at 5 m/s,

$d_2 = \left(\dfrac{10 \text{ m}}{5 \text{ m/s}}\right)\left(\dfrac{1}{2} \text{ m/s}\right) = 1$ m.

In general, for each n, $d_n = \left(\dfrac{d_{n-1}}{5}\right)\left(\dfrac{1}{2}\right)$

$= \dfrac{d_{n-1}}{10}$.

The total distance traveled by the tortoise is

$\sum\limits_{n=1}^{\infty} d_n = 10 + 1 + \dfrac{1}{10} + \dots$ which is a

geometric series with $a = 10$, $r = \dfrac{1}{10}$.

This converges to $\dfrac{10}{1 - \frac{1}{10}} = \dfrac{100}{9}$ m.

Section 12.3   The Integral Test and Estimates of Sums

This section and the next three sections provide tests to determine whether a series converges without explicitly finding the sum. (It is a good first step to determine whether something exists before trying to calculate or estimate it.) The Integral Test in this section is a natural first choice, for it relates infinite series $\left(\sum\limits_{n=1}^{\infty} \ldots \right)$ to improper integrals $\left(\int_{1}^{\infty} \ldots dn \right)$(; however it only applies to certain series of positive terms. The test, when applicable, also permits us to estimate the sum of the convergent series.

Concepts to Master

A. Integral Test for convergence; p-series

B. Estimate of the sum of a convergent series using the Integral Test

Summary and Focus Questions

Page 760 (ET Page 724)

A. *Integral Test*: Let $\{a_n\}$ be a sequence of positive, decreasing terms $(a_{n+1} \le a_n$ for all $n)$. Suppose $f(x)$ is a positive, continuous, and decreasing function on $[1, \infty)$ such that $a_n = f(n)$ for all n. Then $\sum\limits_{n=1}^{\infty} a_n$ converges if and only if $\int_{1}^{\infty} f(x)dx$ converges.

The Integral Test works well if a_n has the form of a function whose antiderivative is easily found. It is one of the few general tests that gives necessary and sufficient conditions for convergence.

A *p-series* has the form $\sum\limits_{n=1}^{\infty} \dfrac{1}{n^p}$, where p is a constant. A p-series converges (by the Integral Test) if and only if $p > 1$.

1) True or False:
The Integral Test will determine to what value a series converges.

False. The test only indicates whether a series converges.

2) Test $\displaystyle\sum_{n=1}^{\infty} \frac{n}{e^n}$ for convergence.

$\dfrac{n}{e^n}$ suggests the function $f(x) = xe^{-x}$.
On the interval $[1, \infty], f$ is continuous and decreasing.

$$\int_1^{\infty} xe^{-x} = \lim_{t\to\infty} \int_1^t xe^{-x}\, dx.$$

Using integration by parts,
$(u = x, \, dv = e^{-x}\, dx)$

$$\int_1^t xe^{-x}\, dx = -xe^{-x} - e^{-x}\Big]_1^t = \frac{2}{e} - \frac{t+1}{e^t}.$$

$$\lim_{t\to\infty} \left(\frac{2}{e} - \frac{t+1}{e^t}\right) = \frac{2}{e} - 0 = \frac{2}{e}.$$

(By L'Hospital's Rule, $\dfrac{t+1}{e^t} \to 0$.)

Thus $\displaystyle\int_1^{\infty} xe^{-x}\, dx = \frac{2}{e}$, so $\displaystyle\sum_{n=1}^{\infty} \frac{n}{e^n}$ converges.

Note: We can *not* conclude that $\displaystyle\sum_{n=1}^{\infty} \frac{n}{e^n} = \frac{2}{e}$.

3) Which of these converge?

a) $\displaystyle\sum_{n=1}^{\infty} \frac{1}{n^2}$.

Converges (*p*-series with $p = 2$).

b) $\displaystyle\sum_{n=1}^{\infty} \frac{1}{\sqrt{n}}$.

Diverges $\left(p\text{-series with } p = \frac{1}{2}\right)$.

c) $\displaystyle\sum_{n=1}^{\infty} \frac{3}{2n^3}$.

Converges. $\displaystyle\sum_{n=1}^{\infty} \frac{3}{2n^3} = \frac{3}{2}\sum_{n=1}^{\infty} \frac{1}{n^3}$ is a multiple of a *p*-series with $p = 3$.

d) $\displaystyle\sum_{n=1}^{\infty} \left(\frac{1}{n^3} + \frac{1}{8^n}\right)$.

Converges.

$$\sum_{n=1}^{\infty} \left(\frac{1}{n^3} + \frac{1}{8^n}\right) = \sum_{n=1}^{\infty} \frac{1}{n^3} + \sum_{n=1}^{\infty} \frac{1}{8^n},$$

the sum of a convergent *p*-series ($p = 3$) and a convergent geometric series $\left(r = \frac{1}{8}\right)$.

Page 762 (ET Page 726)

B. If $\displaystyle\sum_{n=1}^{\infty} a_n = s$ has been found to converge by the Integral Test using $f(x)$, then the error $R_n = s - s_n$ between the series value and the nth partial sum satisfies

$$\int_{n+1}^{\infty} f(x)\, dx \le s - s_n \le \int_n^{\infty} f(x)\, dx.$$

4) a) Find the sixth partial sum of $\sum\limits_{n=1}^{\infty} \frac{1}{n^2}$.

b) Estimate the difference between your answer to part a) and the exact value of
$$s = \sum\limits_{n=1}^{\infty} \frac{1}{n^2}.$$

c) Estimate $s = \sum\limits_{n=1}^{\infty} \frac{1}{n^2}$ using the results of parts a) and b).

5) How many terms of $\sum\limits_{n=1}^{\infty} \frac{1}{n^2+1}$ are necessary to find the sum to within 0.01?

$s_6 = 1 + \frac{1}{4} + \frac{1}{9} + \frac{1}{16} + \frac{1}{25} + \frac{1}{36}$

$= \frac{5369}{3600} \approx 1.491.$

We know by the Integral Test using

$f(x) = \frac{1}{x^2}$ that $\sum\limits_{n=1}^{\infty} \frac{1}{n^2}$ converges. Thus

$\int_7^{\infty} \frac{1}{x^2}\, dx \le s - s_6 \le \int_6^{\infty} \frac{1}{x^2}\, dx.$

$\int_7^{\infty} \frac{1}{x^2}\, dx = \lim\limits_{t\to\infty} \int_7^t x^{-2}\, dx$

$= \lim\limits_{t\to\infty} \left(\frac{1}{7} - \frac{1}{t}\right) = \frac{1}{7}.$

Likewise $\int_6^{\infty} \frac{1}{x^2}\, dx = \frac{1}{6}.$

Therefore $\frac{1}{7} \le s - s_6 \le \frac{1}{6}.$

$\frac{1}{7} \le s - s_6 \le \frac{1}{6}.$

$\frac{1}{7} + s_6 \le s \le \frac{1}{6} + s_6.$

$\frac{1}{7} + \frac{5369}{3600} \le s \le \frac{1}{6} + \frac{5369}{3600}$

$\frac{41183}{25200} \le s \le \frac{5969}{3600}$

$1.634 \le s \le 1.658$

It turns out that $s = \frac{\pi^2}{6} \approx 1.645.$

Let $f(x) = \frac{1}{x^2 + 1}.$

$\sum\limits_{n=1}^{\infty} \frac{1}{n^2 + 1}$ converges because $\int_1^{\infty} \frac{1}{x^2 + 1}\, dx$

$= \lim\limits_{t\to\infty} \int_1^t \frac{1}{x^2 + 1}\, dx$

$= \lim\limits_{t\to\infty} (\tan^{-1}(t) - \tan^{-1}(1)) = \frac{\pi}{2} - \frac{\pi}{4} = \frac{\pi}{4}.$

Since $s - s_n \le \int_n^{\infty} f(x)\, dx$, it is sufficient to

find n such that $\int_n^{\infty} f(x)\, dx \le 0.01.$

From above, we see that

$\int_n^{\infty} \frac{1}{x^2 + 1}\, dx = \frac{\pi}{2} - \tan^{-1}(n).$

$\frac{\pi}{2} - \tan^{-1}(n) < 0.01$

$\tan^{-1}(n) > \frac{\pi}{2} - 0.01$

$n > \tan\left(\frac{\pi}{2} - 0.01\right) \approx 99.997.$

The first 100 terms are sufficient.

Section 12.4   The Comparison Tests

This section deals only with series of positive terms. If a series with positive terms converges, then any other series that is term for term smaller than that one will also converge. This section makes this notion of comparing series of positive terms precise and discusses estimating limits.

Concepts to Master

A. Comparison Test; Limit Comparison Test

B. Estimate the sum of a convergent series

Summary and Focus Questions

Page 767 (ET Page 731)

A. Let $\displaystyle\sum_{n=1}^{\infty} a_n$ and $\displaystyle\sum_{n=1}^{\infty} b_n$ be series whose terms are all positive.

Comparison Test:

1. If $a_n \le b_n$ for all n and $\displaystyle\sum_{n=1}^{\infty} b_n$ converges, then $\displaystyle\sum_{n=1}^{\infty} a_n$ converges.

2. If $a_n \ge b_n$ for all n and $\displaystyle\sum_{n=1}^{\infty} b_n$ diverges, then $\displaystyle\sum_{n=1}^{\infty} a_n$ diverges.

Limit Comparison Test:

1. If $\displaystyle\lim_{n\to\infty} \frac{a_n}{b_n} = c > 0$, then $\displaystyle\sum_{n=1}^{\infty} a_n$ and $\displaystyle\sum_{n=1}^{\infty} b_n$ either both converge or both diverge.

2. If $\displaystyle\lim_{n\to\infty} \frac{a_n}{b_n} = 0$ and $\displaystyle\sum_{n=1}^{\infty} b_n$ converges, then $\displaystyle\sum_{n=1}^{\infty} a_n$ converges.

3. If $\displaystyle\lim_{n\to\infty} \frac{a_n}{b_n} = 0$ and $\displaystyle\sum_{n=1}^{\infty} a_n$ diverges, then $\displaystyle\sum_{n=1}^{\infty} b_n$ diverges.

For a given series $\displaystyle\sum_{n=1}^{\infty} a_n$, success using either comparison test to determine whether $\displaystyle\sum_{n=1}^{\infty} a_n$ converges will depend on coming up with another series $\displaystyle\sum_{n=1}^{\infty} b_n$ (geometric, *p*-series, ...) whose convergence is known and for which a_n and b_n may be compared.

Example: Determine whether $\sum_{n=1}^{\infty} \frac{1}{\sqrt{n}+4}$ converges.

You need a hunch beforehand whether $\sum_{n=1}^{\infty} \frac{1}{\sqrt{n}+4}$ converges. Because

$\frac{1}{\sqrt{n}+4}$ is "like" $\frac{1}{\sqrt{n}}$ and $\sum_{n=1}^{\infty} \frac{1}{\sqrt{n}}$ is a divergent p-series $\left(p = \frac{1}{2}\right)$, we have

reason to believe that $\sum_{n=1}^{\infty} \frac{1}{\sqrt{n}+4}$ diverges. We also have an idea of the type

of b_n to look for—one for which $a_n \geq b_n$:

Let $b_n = \frac{1}{3\sqrt{n}}$. Then for all n, $\frac{1}{\sqrt{n}+4} \geq \frac{1}{3\sqrt{n}}$. $\sum_{n=1}^{\infty} \frac{1}{3\sqrt{n}} = \frac{1}{3} \sum_{n=1}^{\infty} n^{-\frac{1}{2}}$

is a divergent p-series. Therefore, $\sum_{n=1}^{\infty} \frac{1}{\sqrt{n}+4}$ diverges.

---

1) Determine whether or not each of the following converge. Find a series $\sum_{n=1}^{\infty} b_n$ to use with the Comparison Test or Limit Comparison Test.

a) $\sum_{n=1}^{\infty} \frac{1}{n^2+2n}$, $b_n = $ _____.

Because $\frac{1}{n^2+2n}$ is "like" $\frac{1}{n^2}$ for large n, and $\sum_{n=1}^{\infty} \frac{1}{n^2}$ is a convergent p-series, we suspect that the given series converges.

Use $b_n = \frac{1}{n^2}$. Since $\frac{1}{n^2+2n} < \frac{1}{n^2}$ and $\sum_{n=1}^{\infty} \frac{1}{n^2}$ converges, $\sum_{n=1}^{\infty} \frac{1}{n^2+2n}$ converges by the Comparison Test.

b) $\sum_{n=1}^{\infty} \frac{\sqrt[3]{n}}{n+4}$, $b_n = $ _____.

Since $\frac{\sqrt[3]{n}}{n+4}$ is "like" $\frac{\sqrt[3]{n}}{n} = \frac{1}{\sqrt[3]{n^2}}$,

use $b_n = \frac{1}{\sqrt[3]{n^2}}$.

Then $\lim_{n\to\infty} \frac{a_n}{b_n} = \lim_{n\to\infty} \frac{\frac{\sqrt[3]{n}}{n+4}}{\frac{1}{\sqrt[3]{n^2}}} = \lim_{n\to\infty} \frac{n}{n+4} = 1.$

Since $\sum_{n=1}^{\infty} \frac{1}{\sqrt[3]{n^2}}$ diverges $\left(\text{a } p\text{-series with}\right.$

$\left. p = \frac{2}{3}\right)$, $\sum_{n=1}^{\infty} \frac{\sqrt[3]{n}}{n+4}$ diverges by the Limit Comparison Test.

c) $\sum\limits_{n=1}^{\infty} \dfrac{1}{n+2^n}$, $b_n =$ _____.

Use $b_n = \dfrac{1}{2^n}$. Then for $n \geq 1$, $\dfrac{1}{n+2^n} \leq \dfrac{1}{2^n}$.

Since $\sum\limits_{n=1}^{\infty} \dfrac{1}{2^n}$ converges $\left(\text{a geometric series}\right.$

with $r = \dfrac{1}{2}\Big)$, $\sum\limits_{n=1}^{\infty} \dfrac{1}{n+2^n}$ converges.

2) Suppose $\{a_n\}$ and $\{b_n\}$ are positive sequences. Sometimes, Always, Never:

a) If $\lim\limits_{n\to\infty} \dfrac{a_n}{b_n} = 0$ and $\sum\limits_{n=1}^{\infty} a_n$ converges,

then $\sum\limits_{n=1}^{\infty} b_n$ converges.

Sometimes. True for $a_n = \dfrac{1}{n^3}$ and $b_n = \dfrac{1}{n^2}$.
False for $a_n = \dfrac{1}{n^2}$ and $b_n = \dfrac{1}{n}$.

b) If $\lim\limits_{n\to\infty} \dfrac{a_n}{b_n} = \infty$ and $\sum\limits_{n=1}^{\infty} b_n$ diverges,

then $\sum\limits_{n=1}^{\infty} a_n$ diverges.

Always.

Page 769 (ET Page 733)

B. If $\sum\limits_{n=1}^{\infty} a_n = s$ converges by the Comparison Test using $t = \sum\limits_{n=1}^{\infty} b_n$, then

$$s - s_n \leq t - t_n.$$

This means that $t - t_n$ (which may be easier to calculate) is an upper estimate for $s - s_n$.

3) In question 1c), $\sum\limits_{n=1}^{\infty} \dfrac{1}{n+2^n}$ converges.

a) Find s_4, the fourth partial sum.

$$s_4 = \dfrac{1}{1+2} + \dfrac{1}{2+4} + \dfrac{1}{3+8} + \dfrac{1}{4+16}$$
$$= \dfrac{1}{3} + \dfrac{1}{6} + \dfrac{1}{11} + \dfrac{1}{20} = \dfrac{141}{220}.$$

b) Estimate the difference between this series and its fourth partial sum.

From 1c), $b_n = \dfrac{1}{2^n}$ may be used to show

$\sum \dfrac{1}{n+2^n}$ converges. Let $s = \sum\limits_{n=1}^{\infty} \dfrac{1}{n+2^n}$ and

$t = \sum\limits_{n=1}^{\infty} \dfrac{1}{2^n}$. Then $s - s_4 \leq t - t_4$.

Since t is the result of a geometric series we can calculate it:

$$t = \dfrac{\dfrac{1}{2}}{1 - \dfrac{1}{2}} = 1$$

$$t_4 = \dfrac{1}{2} + \dfrac{1}{4} + \dfrac{1}{8} + \dfrac{1}{16} = \dfrac{15}{16}.$$

Thus, $t - t_4 = 1 - \dfrac{15}{16} = \dfrac{1}{16}$ and

$s - s_4 \leq \dfrac{1}{16}$. Therefore, we know $s_4\left(\dfrac{141}{220}\right)$ is

within $\dfrac{1}{16}$ of the value of s.

Section 12.5 Alternating Series

An alternating series is one for which consecutive terms have opposite signs. This section shows you a test for convergence of these kinds of series and an especially simple method to estimate the sum of a convergent alternating series.

Concepts to Master

A. Alternating Series Test

B. Estimating the sum of a convergent alternating series

Summary and Focus Questions

Page 772 (ET Page 736)

A. An *alternating series* has successive terms of opposite sign—that is, it has one of these forms:

$$\sum_{n=1}^{\infty} (-1)^{n-1} a_n \quad \text{or} \quad \sum_{n=1}^{\infty} (-1)^n a_n, \text{ where } a_n > 0.$$

The Alternating Series Test:

If $\{a_n\}$ is a decreasing sequence with $\lim_{n\to\infty} a_n = 0$, then $\sum_{n=1}^{\infty} (-1)^{n-1} a_n$ converges.

1) Is $\sum_{n=1}^{\infty} \frac{\sin n}{n}$ an alternating series?

No, although some terms are positive and others negative.

2) Determine whether each converge:

a) $\sum_{n=1}^{\infty} \frac{(-1)^{n-1}}{e^n}$.

This is an alternating series with $a_n = \frac{1}{e^n}$.

$\frac{1}{e^{n+1}} \leq \frac{1}{e^n}$, so the terms decrease.

Since $\lim_{n\to\infty} \frac{1}{e^n} = 0$, $\sum_{n=1}^{\infty} \frac{(-1)^n}{e^n}$ converges by the Alternating Series Test.

b) $\displaystyle\sum_{n=1}^{\infty} \frac{(-1)^n n}{n+1}.$

This is an alternating series but the other conditions for the test do not hold.

Since $\displaystyle\lim_{n\to\infty} \frac{n}{n+1} \neq 0$, $\displaystyle\sum_{n=1}^{\infty} \frac{(-1)^n n}{n+1}$ diverges by the Divergence Test.

c) $\displaystyle\sum_{n=1}^{\infty} \frac{(-1)^{n-1} \ln n}{n}$

It may not be obvious that $a_n = \dfrac{\ln n}{n}$ decreases. Let $f(x) = \dfrac{\ln x}{x}$.

Then $f'(x) = \dfrac{x\left(\frac{1}{x}\right) - \ln x\,(1)}{x^2} = \dfrac{1 - \ln x}{x^2}.$

For $x > 1, f'(x) < 0$. Thus $f(x)$, and hence $\{a_n\}$ is decreasing. Finally,

$$\lim_{n\to\infty} \frac{\ln n}{n} = \lim_{n\to\infty} \frac{\frac{1}{n}}{1} = 0 \text{ by L'Hospital's Rule.}$$

Therefore, $\displaystyle\sum_{n=1}^{\infty} \frac{(-1)^{n-1} \ln n}{n}$ converges.

B. If $\displaystyle\sum_{n=1}^{\infty} (-1)^{n-1} a_n$ is an alternating series with $0 \le a_{n+1} \le a_n$ and

Page 774 (ET Page 738)

$\displaystyle\lim_{n\to\infty} a_n = 0$ then

$$\left| s_n - s \right| \le a_{n+1}.$$

Since the difference between the limit of the series and the nth partial sum does not exceed a_{n+1}, s_n may be used to approximate s to within a_{n+1}.

3) The series $\displaystyle\sum_{n=1}^{\infty} \frac{(-1)^n}{\sqrt{n}+1}$ converges.

Estimate the error between the sum of the series and its fifteenth partial sum.

Since the series alternates and $\dfrac{1}{\sqrt{n}+1}$ decreases to 0, the error $\left| s_{15} - s \right|$ does not exceed $a_{16} = \dfrac{1}{\sqrt{16}+1} = 0.2.$

4) For what value of n is s_n, the nth partial sum, within 0.001 of $s = \sum_{n=1}^{\infty} \frac{(-1)^{n+1}}{2n+5}$?

$$|s_n - s| \le a_{n+1} = \frac{1}{2(n+1)+5} = \frac{1}{2n+7}.$$

$$\frac{1}{2n+7} \le 0.001.$$

$$2n + 7 \ge 1000,$$

$$2n \ge 993,$$

$$n \ge 496.5.$$

Let $n = 497$. Then s_{497} is within 0.001 of s.

5) Approximate $\sum_{n=1}^{\infty} \frac{(-1)^{n+1}}{n^3}$ to within 0.005.

$$|s_n - s| \le a_{n+1} = \frac{1}{(n+1)^3} \le 0.005 = \frac{1}{200}.$$

$$(n+1)^3 \ge 200$$

$$n + 1 \ge \sqrt[3]{200} \approx 5.84$$

$$n \ge 4.84. \text{ Let } n = 5.$$

Therefore s_5 is within 0.005 of s.

$$s_5 = \frac{1}{1^3} - \frac{1}{2^3} + \frac{1}{3^3} - \frac{1}{4^3} + \frac{1}{5^3}$$

$$= 1 - \frac{1}{8} + \frac{1}{27} - \frac{1}{64} + \frac{1}{125} \approx 0.9044.$$

We do not know to what the series converges, but that value is within 0.005 of 0.9044.

Section 12.6 Absolute Convergence and the Ratio and Root Tests

The concept of convergence of an infinite series may be split into two separate subconcepts: absolute convergence and conditional convergence. This section gives a precise definition of each. It also describes the Root and Ratio Tests, which may be used to check for absolute convergence.

Concepts to Master

A. Absolute convergence; Conditional convergence

B. Ratio Test; Root Test

C. Rearrangement of terms of a series

Summary and Focus Questions

Page 776 (ET Page 740)

A. For any series (with a_n not necessarily positive or alternating) two types of convergence may be defined:

$$\sum_{k=1}^{\infty} a_k \text{ converges } \textit{absolutely} \text{ means that } \sum_{k=1}^{\infty} |a_k| \text{ converges.}$$

$$\sum_{k=1}^{\infty} a_k \text{ converges } \textit{conditionally} \text{ means that } \sum_{k=1}^{\infty} a_k \text{ converges but } \sum_{k=1}^{\infty} |a_k| \text{ diverges.}$$

Either one of absolute convergence or conditional convergence implies (ordinary) convergence. Conversely, convergence implies either absolute or conditional convergence. Thus *every* series must behave in exactly one of these three ways: diverge, converge absolutely, or converge conditionally.

Examples: $\sum_{n=1}^{\infty} \dfrac{(-1)^{n-1}}{n^3}$ is absolutely convergent because the series

$\sum_{n=1}^{\infty} \left| \dfrac{(-1)^{n-1}}{n^s} \right| = \sum_{n=1}^{\infty} \dfrac{1}{n^3}$ converges (*p*-series, $p = 3$). On the other hand,

$\sum_{n=1}^{\infty} \dfrac{(-1)^{n-1}}{\sqrt[3]{n}}$ is conditionally convergent because it converges (by the

Alternating Series Test) but $\sum_{n=1}^{\infty} \left| \dfrac{(-1)^{n-1}}{\sqrt[3]{n}} \right| = \sum_{n=1}^{\infty} \dfrac{1}{\sqrt[3]{n}}$ diverges $\left(p\text{-series}, p = \dfrac{1}{3} \right)$.

1) Sometimes, Always, or Never:

a) If $\displaystyle\sum_{n=1}^{\infty} |a_n|$ diverges, then $\displaystyle\sum_{n=1}^{\infty} a_n$ diverges.

Sometimes. True for $a_n = n$ but false for $a_n = \dfrac{(-1)^n}{n}$.

b) If $\displaystyle\sum_{n=1}^{\infty} a_n$ diverges, then $\displaystyle\sum_{n=1}^{\infty} |a_n|$ diverges.

Always.

c) If $a_n \geq 0$ for all n, then $\displaystyle\sum_{n=1}^{\infty} a_n$ is not conditionally convergent.

Always, since $|a_n| = a_n$ for all n.

2) Determine whether each converges conditionally, converges absolutely, or diverges.

a) $\displaystyle\sum_{n=1}^{\infty} \frac{(-1)^n}{\sqrt{n}}$.

The series is alternating with $\dfrac{1}{\sqrt{n}}$ decreasing to 0 so it converges. It remains to check for absolute convergence:

$\displaystyle\sum_{n=1}^{\infty} \left|\frac{(-1)^n}{\sqrt{n}}\right| = \sum_{n=1}^{\infty} \frac{1}{\sqrt{n}}$ diverges (p-series with $p = \frac{1}{2}$). Thus $\displaystyle\sum_{n=1}^{\infty} \frac{(-1)^n}{\sqrt{n}}$ converges conditionally.

b) $\displaystyle\sum_{n=1}^{\infty} \frac{\sin n + \cos n}{n^3}$.

Check for absolute convergence first:
$\displaystyle\sum_{n=1}^{\infty} \left|\frac{\sin n + \cos n}{n^3}\right| = \sum_{n=1}^{\infty} \frac{|\sin n + \cos n|}{n^3}$.
Since $|\sin n + \cos n| \leq 2$ and $\displaystyle\sum_{n=1}^{\infty} \frac{2}{n^3}$ converges (it is a p-series with $p = 3$), $\displaystyle\sum_{n=1}^{\infty} \frac{|\sin n + \cos n|}{n^3}$ converges by the Comparison Test. Thus $\displaystyle\sum_{n=1}^{\infty} \frac{\sin n + \cos n}{n^3}$ converges absolutely.

3) Does $\displaystyle\sum_{n=1}^{\infty} \frac{\sin e^n}{e^n}$ converge?

Yes. The series is not alternating but does contain both positive and negative terms. Check for absolute convergence:

$$\sum_{n=1}^{\infty} \left|\frac{\sin e^n}{e^n}\right| = \sum_{n=1}^{\infty} \frac{|\sin e^n|}{e^n}.$$

Since $|\sin e^n| \le 1$, $\dfrac{|\sin e^n|}{e^n} \le \dfrac{1}{e^n}$.

$\displaystyle\sum_{n=1}^{\infty} \frac{1}{e^n}$ converges $\left(\text{geometric series with } r = \frac{1}{e}\right)$ so by the Comparison Test $\displaystyle\sum_{n=1}^{\infty} \left|\frac{\sin e^n}{e^n}\right|$ converges. Therefore $\displaystyle\sum_{n=1}^{\infty} \frac{\sin e^n}{e^n}$ converges absolutely and hence converges.

Page
778
(ET Page
742)

B. The following tests are very useful for determining whether a series converges absolutely. Neither involves another series or function, which makes them relatively easy to use; however, each has cases where it fails.

The Ratio Test works well when the nth term, a_n, contains exponentials or factorials or when a_n is defined recursively. It will fail when a_n is a rational function of n.

Ratio Test: Let a_n be a sequence of nonzero terms.

1. If $\displaystyle\lim_{n\to\infty} \left|\frac{a_{n+1}}{a_n}\right| = L < 1$, then $\displaystyle\sum_{n=1}^{\infty} a_n$ converges absolutely.

2. If $\displaystyle\lim_{n\to\infty} \left|\frac{a_{n+1}}{a_n}\right| = L > 1$ or $\displaystyle\lim_{n\to\infty} \left|\frac{a_{n+1}}{a_n}\right| = \infty$, then $\displaystyle\sum_{n=1}^{\infty} a_n$ diverges.

3. If $\displaystyle\lim_{n\to\infty} \left|\frac{a_{n+1}}{a_n}\right| = 1$, the Ratio Test fails: $\displaystyle\sum_{n=1}^{\infty} a_n$ may converge or diverge.

The Root Test works well when a_n contains an expression to the nth power.

Root Test:

1. If $\displaystyle\lim_{n\to\infty} \sqrt[n]{|a_n|} = L < 1$, then $\displaystyle\sum_{n=1}^{\infty} a_n$ converges absolutely.

2. If $\displaystyle\lim_{n\to\infty} \sqrt[n]{|a_n|} = L > 1$ or $\displaystyle\lim_{n\to\infty} \sqrt[n]{|a_n|} = \infty$, $\displaystyle\sum_{n=1}^{\infty} a_n$ diverges.

3. If $\displaystyle\lim_{n\to\infty} \sqrt[n]{|a_n|} = 1$, the Root Test fails: $\displaystyle\sum_{n=1}^{\infty} a_n$ may converge or diverge.

4) Determine whether each converges by the Ratio Test.

a) $\sum_{n=1}^{\infty} \frac{n}{4^n}$.

$$\lim_{n\to\infty}\left|\frac{a_{n+1}}{a_n}\right| = \lim_{n\to\infty}\frac{\frac{n+1}{4^{n+1}}}{\frac{n}{4^n}} = \lim_{n\to\infty}\frac{n+1}{4n} = \frac{1}{4}.$$

Thus $\sum_{n=1}^{\infty} \frac{n}{4^n}$ converges absolutely and therefore converges.

b) $\sum_{n=1}^{\infty} \frac{(-4)^n}{n!}$.

$$\lim_{n\to\infty}\left|\frac{a_{n+1}}{a_n}\right| = \lim_{n\to\infty}\left|\frac{\frac{(-4)^{n+1}}{(n+1)!}}{\frac{(-4)^n}{n!}}\right| = \lim_{n\to\infty}\frac{4}{n+1} = 0.$$

Thus $\sum_{n=1}^{\infty} \frac{(-4)^n}{n!}$ converges absolutely and therefore converges.

c) $\sum_{n=1}^{\infty} \frac{(-1)^n}{\sqrt[3]{n}}$.

$$\lim_{n\to\infty}\left|\frac{a_{n+1}}{a_n}\right| = \lim_{n\to\infty}\left|\frac{\frac{(-1)^{n+1}}{\sqrt[3]{n+1}}}{\frac{(-1)^n}{\sqrt[3]{n}}}\right|$$

$$= \lim_{n\to\infty}\sqrt[3]{\frac{n}{n+1}} = 1.$$

The Ratio Test fails. The Alternating Series Test can be used to show that this series converges.

d) $\sum_{n=1}^{\infty} a_n$, where $a_1 = 4$ and
$a_{n+1} = \frac{3a_n}{2n+1}$.

$$\lim_{n\to\infty}\left|\frac{a_{n+1}}{a_n}\right| = \lim_{n\to\infty}\frac{\frac{3a_n}{2n+1}}{a_n} = \lim_{n\to\infty}\frac{3}{2n+1} = 0.$$

Thus $\sum_{n=1}^{\infty} a_n$ converges absolutely and therefore converges.

5) Determine whether each converges by the Root Test:

a) $\displaystyle\sum_{n=1}^{\infty} \frac{1}{(n+1)^n}.$

$$\lim_{n\to\infty} \sqrt[n]{|a_n|} = \lim_{n\to\infty} \sqrt[n]{\frac{1}{(n+1)^n}} = \lim_{n\to\infty} \frac{1}{n+1}$$

$$= 0.$$

Thus $\displaystyle\sum_{n=1}^{\infty} \frac{1}{(n+1)^n}$ converges absolutely and therefore converges.

b) $\displaystyle\sum_{n=1}^{\infty} \frac{3^n}{n^3}.$

$$\lim_{n\to\infty} \sqrt[n]{|a_n|} = \lim_{n\to\infty} \sqrt[n]{\frac{3^n}{n^3}} = \lim_{n\to\infty} \frac{3}{n^{3/n}} = \frac{3}{1} = 3.$$

Therefore by the Root Test $\displaystyle\sum_{n=1}^{\infty} \frac{3^n}{n^3}$ diverges.

To show $\displaystyle\lim_{n\to\infty} n^{3/n} = 1$ let $y = \lim_{n\to\infty} n^{3/n}$.

Then $\ln y = \displaystyle\lim_{n\to\infty} \ln n^{3/n} = \lim_{n\to\infty} \frac{3\ln n}{n}$

$$\left(\text{form } \frac{\infty}{\infty}\right) = \lim_{n\to\infty} \frac{\frac{3}{n}}{1} = 0.$$

Thus $y = e^0 = 1$.

c) $\displaystyle\sum_{n=1}^{\infty} \frac{(-1)^n}{n}$

$$\lim_{n\to\infty} \sqrt[n]{\left|\frac{(-1)^n}{n}\right|} = \lim_{n\to\infty} \frac{1}{\sqrt[n]{n}} = 1.$$

The Root Test fails, but we know this series converges because it is the alternating harmonic series.

Page 781 (ET Page 745)

C. A *rearrangement* of an infinite series is another series obtained from the series by changing the order of the terms.

If $\displaystyle\sum_{n=1}^{\infty} a_n$ converges absolutely, then all rearrangements converge to the same sum.

If $\displaystyle\sum_{n=1}^{\infty} a_n$ converges conditionally, then rearrangements do not all converge to the same sum.

6) Do all rearrangements of $\displaystyle\sum_{n=1}^{\infty} \frac{(-1)^{n-1}}{n^4}$ converge to the same value?

Yes. The series is absolutely convergent since $\displaystyle\sum_{n=1}^{\infty} \frac{1}{n^4}$ is a p-series ($p = 4$).

Section 12.7   Strategy for Testing Series

Page
783
(ET Page
747)

There are no new techniques in this section – just a summary of the strategies to follow for determining whether a given series converges.

Concepts to Master

Apply the various tests of convergence

Summary and Focus Questions

The main strategy for determining whether a given series converges is to observe the form of the nth term—whether it matches a certain type (p-series, alternating, etc.), resembles a known form (comparison tests or Integral Test), or is amenable to calculation using the Ratio or Root Tests. Here is a brief summary of each test together with a representative example of its use:

Test Name	Form/Conditions	Conclusion(s)	Example
Divergence	$\lim\limits_{n\to\infty} a_n \neq 0$	$\sum a_n$ diverges	$\sum\limits_{n=1}^{\infty} \dfrac{n}{n+1}$
Integral	Find $f(x)$, decreasing with $f(n) = a_n \geq 0$.	$\sum a_n$ converges if and only if $\int_1^{\infty} f(x)\,dx$ converges.	$\sum\limits_{n=1}^{\infty} \dfrac{\ln n}{n}$ $\int \dfrac{\ln x}{x}\,dx = \dfrac{(\ln x)^2}{2}$
p-series	$a_n = \dfrac{1}{n^p}$	$\sum \dfrac{1}{n^p}$ converges if and only if $p > 1$.	$\sum\limits_{n=1}^{\infty} \dfrac{1}{n^3}$
Geometric Series	$a_n = ar^{n-1}$	$\sum ar^{n-1}$ converges if and only if $|r| < 1$.	$\sum\limits_{n=1}^{\infty} \dfrac{2}{3^n}$
Alternating Series	$a_n = (-1)^n b_n$ $(b_n \geq 0)$	$\sum a_n$ converges if $\lim\limits_{n\to\infty} b_n = 0$ and b_n is decreasing.	$\sum\limits_{n=1}^{\infty} \dfrac{(-1)^n}{n}$
Comparison	$0 \leq a_n \leq b_n$	If $\sum b_n$ converges then $\sum a_n$ converges. If $\sum a_n$ diverges then $\sum b_n$ diverges.	$\sum\limits_{n=1}^{\infty} \dfrac{1}{n^3 + 1}$ Choose $b_n = \dfrac{1}{n^3}$.
Limit Comparison	$\lim\limits_{n\to\infty} \dfrac{a_n}{b_n} = c$	If $0 < c < \infty$, $\sum a_n$ converges if and only if $\sum b_n$ does.	$\sum\limits_{n=1}^{\infty} \dfrac{1}{2n^3 + 1}$ Choose $b_n = \dfrac{1}{n^3}$.

Test Name	Form/Conditions	Conclusion(s)	Example
Ratio	$\lim\limits_{n \to \infty} \left| \dfrac{a_{n+1}}{a_n} \right| = L$	If $L < 1$, $\sum a_n$ converges absolutely. If $L > 1$, $\sum a_n$ diverges.	$\sum\limits_{n=1}^{\infty} \dfrac{n^2}{3^n}$
Root	$\lim\limits_{n \to \infty} \sqrt[n]{\lvert a_n \rvert} = L$	If $L < 1$, $\sum a_n$ converges absolutely. If $L > 1$, $\sum a_n$ diverges.	$\sum\limits_{n=1}^{\infty} \left(\dfrac{2n+1}{3n+4} \right)^n$

For each series, what is an appropriate test to apply for each?

1) $\sum\limits_{n=1}^{\infty} \dfrac{(n+1)^n}{4^n}$.

Root Test, because the $(n+1)^n$ and 4^n terms have n in the exponent.

2) $\sum\limits_{n=1}^{\infty} \dfrac{5}{n^2 + 6n + 3}$.

Comparison Test, compare to $\sum \dfrac{1}{n^2}$.

3) $\sum\limits_{n=1}^{\infty} \dfrac{(-1)^n n}{n^6 + 1}$.

Alternating Series Test for convergence, then Comparison Test (compare to $\sum \dfrac{1}{n^5}$) for absolute convergence.

4) $\sum\limits_{n=1}^{\infty} \dfrac{2^n n^3}{n!}$.

Ratio Test, because of the $n!$ term.

5) $\sum\limits_{n=1}^{\infty} \dfrac{n}{e^{n^2}}$

The exponential e^{n^2} suggests the Ratio Test; the function $f(x) = \dfrac{x}{e^{x^2}} = xe^{-x^2}$ may be used in the Integral Test.

Section 12.8   Power Series

This section introduces a special type of function called a power series whose functional values are infinite series. As such, there will be values for its variable x for which the power series is defined (for that x, the resulting series converges) and others for which it does not exist (resulting series diverges). We shall see that the tests of convergence will help in determining the domain of a power series.

Concepts to Master

Power series; Interval of Convergence; Radius of convergence

Summary and Focus Questions

Page 785 (ET Page 749)

A *power series in* $(x\text{-}a)$ is an expression of the form

$$\sum_{n=0}^{\infty} c_n (x - a)^n, \text{ where } c_n \text{ and } a \text{ are constants.}$$

For any particular value of x, the value of the power series is an infinite series. Thus, for some values of x the expression converges and for others it may diverge.

The domain of a power series in $(x\text{-}a)$ is those values of x for which it converges and is called the *interval of convergence*. It consists of all real numbers from $a - R$ to $a + R$ for some number R, where $0 \leq R \leq \infty$. The power series diverges for all x outside the interval of convergence. The number a is called the *center* and R is the *radius* of convergence. When $R = 0$, the interval is the single point $\{a\}$; when $R = \infty$, the interval is all real numbers.

The value of R is found by applying the Ratio Test to $\sum_{n=0}^{\infty} c_n (x - a)^n$ and solving for $|x - a|$ in the resulting inequality:

$$\lim_{n \to \infty} \left| \frac{c_{n+1}(x - a)^{n+1}}{c_n(x - a)^n} \right| = \lim_{n \to \infty} \left| \frac{c_{n+1}}{c_n} \right| |x - a| < 1, \text{ for } |x - a|.$$

When $0 < R < \infty$, the two endpoints $a - R$ and $a + R$ pose special problems: the Ratio Test fails and the endpoint may or may not be in the interval of convergence. You must use some other test to determine convergence at each endpoint.

1) True or False:
A power series is an infinite series.

False. A power series is an expression for a function, $f(x)$. For any x in the domain of f, $f(x)$ is the sum of a convergent series.

2) For what value of x does $\sum_{n=0}^{\infty} c_n(x-a)^n$ always converge?

At $x = a$, $\sum_{n=0}^{\infty} c_n(x-a)^n = c_0$.

The power series always converges when x is the center.

3) Compute the value of:

a) $\sum_{n=0}^{\infty} n^2(x-3)^n$ at $x = 4$.

When $x = 4$ the power series is

$$\sum_{n=0}^{\infty} n^2 1^n = \sum_{n=0}^{\infty} n^2.$$ This infinite series

diverges so there is no value for $x = 4$.

b) $\sum_{n=0}^{\infty} 2^n(x-1)^n$ at $x = \frac{5}{6}$.

When $x = \frac{5}{6}$ the power series is

$$\sum_{n=0}^{\infty} 2^n\left(-\frac{1}{6}\right)^n = \sum_{n=0}^{\infty} \left(-\frac{1}{3}\right)^n$$

$$= 1 - \frac{1}{3} + \frac{1}{9} - \frac{1}{27} + \dots$$

This is a geometric series with $a = 1$,

$r = -\frac{1}{3}$. It converges to $\dfrac{1}{1-\left(-\frac{1}{3}\right)} = \dfrac{3}{4}$.

4) Find the center, radius, and interval of convergence for:

a) $\sum_{n=1}^{\infty} (2n)!(x-1)^n$.

Use the Ratio Test:

$$\lim_{n\to\infty} \left| \frac{(2(n+1))!(x-1)^{n+1}}{(2n)!(x-1)^n} \right|$$

$$= \left(\lim_{n\to\infty} (2n+1)(2n+2) \right)|x-1| = \infty,$$

for all x except when $x = 1$. Thus the center is 1, the radius is 0, and the interval of convergence is $\{1\}$.

b) $\displaystyle\sum_{n=1}^{\infty} \frac{(x-4)^n}{2^n n}$.

Use the Ratio Test:

$$\lim_{n\to\infty} \left| \frac{(x-4)^{n+1}}{2^{n+1}(n+1)} \cdot \frac{2^n n}{(x-4)^n} \right|$$

$$= \lim_{n\to\infty} \frac{n}{2(n+1)} |x-4| = \tfrac{1}{2}|x-4|.$$

From $\tfrac{1}{2}|x-4| < 1$ we conclude

$|x-4| < 2$. The center is 4 and radius is 2.

When $|x-4| = 2$, $x = 2$ or $x = 6$.

At $x = 2$, the power series becomes

$\displaystyle\sum_{n=1}^{\infty} \frac{(-1)^n}{n}$ which converges by the

Alternating Series Test.

At $x = 6$, the power series is the divergent

harmonic series $\displaystyle\sum_{n=1}^{\infty} \frac{1}{n}$.

The interval of convergence is $[2, 6)$.

c) $\displaystyle\sum_{n=1}^{\infty} \frac{x^n}{n!}$

Use the Ratio Test:

$$\lim_{n\to\infty} \left| \frac{\frac{x^n+1}{(n+)!}}{\frac{x^n}{n!}} \right| = \lim_{n\to\infty} \frac{|x|}{n+1} = 0$$

for all values of x. Thus the interval of
convergence is all real numbers, $(-\infty, \infty)$.

5) Suppose $\displaystyle\sum_{n=0}^{\infty} c_n (x-5)^n$ converges for

$x = 8$. For what other values must it
converge?

The interval of convergence is $5 - R$ to
$5 + R$.

$$\begin{array}{ccccc} + & + & + & + & + \\ 5-R & 2 & 5 & 8 & 5+R \end{array}$$

Because 8 is in this interval $8 \leq 5 + R$.
Thus $3 \leq R$ which means the interval at
least contains all numbers between $5 - 3$
and $5 + 3$. Thus the power series converges
for at least all $x \in (2, 8]$.

6) True, False: The two power series

$\displaystyle\sum_{n=0}^{\infty} c_n(x-a)^n$ and $\displaystyle\sum_{n=0}^{\infty} 2c_n (x-a)^n$ have

the same interval of convergence.

True.

Section 12.9 Representations of Functions as Power Series

In the last section we saw that a power series may be used to define a function. In this section we shall turn things around and see that for many functions, such as $f(x) = \dfrac{2x}{1 + x^2}$, it is possible to find a power series expression for $f(x)$.

Concepts to Master

A. Obtaining a power series expression for a function by manipulating known series

B. Differentiation and integration of power series

Summary and Focus Questions

Page 790 (ET Page 754)

A. To write a given function as a power series we start with a known power series for a similar function. Then by substitutions (such as x^2 for x, etc.) and multiplication by constants and powers of x, we turn the power series into one for the given function. An example will help:

Example: Find a power series for $f(x) = \dfrac{2x}{1 + x^2}$.

We start with a known series expression for the similar function $\dfrac{1}{1 - x}$:

$$\frac{1}{1 - x} = \sum_{n = 0}^{\infty} x^n$$

Then substitute $-x^2$ for x:

$$\frac{1}{1 + x^2} = \sum_{n = 0}^{\infty} (-x^2)^n = \sum_{n = 0}^{\infty} (-1)^n x^{2n}$$

Then multiply by $2x$:

$$\frac{2x}{1 + x^2} = \sum_{n = 0}^{\infty} (-1)^n (2x) x^{2n} = \sum_{n = 0}^{\infty} (-1)^n 2x^{2n+1}.$$

The interval of convergence for $\dfrac{1}{1 - x} = \displaystyle\sum_{n = 0}^{\infty} x^n$ is $(-1, 1)$. From $|x| < 1$, the substitution $-x^2$ for x yields $\left| -x^2 \right| < 1$, which is still $|x| < 1$. Therefore, the interval of convergence for $\dfrac{2x}{1 + x^2} = \displaystyle\sum_{n = 0}^{\infty} (-1)^n 2x^{2n+1}$ is $(-1, 1)$.

1) Given $\dfrac{1}{1-x} = \displaystyle\sum_{n=0}^{\infty} x^n, |x| < 1$, find a power series expression for:

a) $\dfrac{1}{2-x}$.

$$\frac{1}{2-x} = \frac{1}{2\left(1-\frac{x}{2}\right)} = \frac{1}{2}\left(\frac{1}{1-\frac{x}{2}}\right).$$

Thus $\dfrac{1}{2-x} = \dfrac{1}{2}\displaystyle\sum_{n=0}^{\infty}\left(\frac{x}{2}\right)^n = \displaystyle\sum_{n=0}^{\infty}\frac{1}{2}\cdot\frac{x^n}{2^n}$

$$= \sum_{n=0}^{\infty} \frac{x^n}{2^{n+1}}.$$

From $\left|\dfrac{x}{2}\right| < 1, |x| < 2$, so $(-2, 2)$ is the interval of convergence.

b) $\dfrac{x}{x^3-1}$.

We note that $\dfrac{x}{x^3-1} = -\dfrac{x}{1-x^3}$.

First substitute x^3 for x:

$$\frac{1}{1-x^3} = \sum_{n=0}^{\infty}(x^3)^n = \sum_{n=0}^{\infty} x^{3n}.$$

Now multiply by $-x$:

$$\frac{-x}{1-x^3} = -x\sum_{n=0}^{\infty} x^{3n} = -\sum_{n=0}^{\infty} x^{3n+1}.$$

Originally $|x| < 1$, hence $|x^3| < 1$. Thus the interval of convergence is still $(-1, 1)$.

Page
791
(ET Page
755)

B. If $f(x) = \displaystyle\sum_{n=0}^{\infty} a_n(x-c)^n$ has interval of convergence with radius R then:

1. f is continuous on $(c - R, c + R)$.

2. For all $x \in (c - R, c + R), f$ may be differentiated term by term:

$$f'(x) = \sum_{n=1}^{\infty} na_n(x-c)^{n-1}.$$

3. For all $x \in (c - R, c + R), f$ may be integrated term by term:

$$\int f(x)dx = C + \sum_{n=0}^{\infty} \frac{a_n(x-c)^{n+1}}{n+1}.$$

Both $f'(x)$ and $\int f(x)dx$ have radius of convergence R but the endpoints $c - R$ and $c + R$ must be checked individually.

Term by term integration and differentiation is another technique to help find power series for functions.

Example: To find a power series for $\ln (1 + x^2)$ recall from part A that the series for $\frac{1}{1 - x}$ was used to find that $\frac{2x}{1 + x^2} = \sum_{n = 0}^{\infty} (-1)^n 2x^{2n+1}$.

Thus $\int \frac{2x}{1 + x^2} dx = \int \sum_{n = 0}^{\infty} (-1)^n 2x^{2n+1} dx = \sum_{n = 0}^{\infty} (-1)^n 2 \int x^{2n+1} dx$

$$= \sum_{n = 0}^{\infty} (-1)^n 2 \frac{x^{2n+2}}{2n + 2} = \sum_{n = 0}^{\infty} (-1)^n \frac{x^{2n+2}}{n + 1}.$$

Since $\ln (1 + x^2) = \int \frac{2x}{1 + x^2} dx$, $\ln (1 + x^2) = \sum_{n = 0}^{\infty} (-1)^n \frac{x^{2n+2}}{n + 1}$.

This series has the same interval of convergence $(-1, 1)$ as the series for $\frac{1}{1 - x}$ but we need to check the endpoints. At both $x = 1$ and $x = -1$, $x^{2n+2} = 1$ and the resulting series is $\sum_{n = 0}^{\infty} \frac{(-1)^n}{n+1}$ converges. Therefore, the interval of convergence for $\ln (1 + x^2) = \sum_{n = 0}^{\infty} (-1)^n \frac{x^{2n+2}}{n + 1}$ is $[-1, 1]$.

2) Find $f'(x)$ for

$$f(x) = \sum_{n = 0}^{\infty} 2^n(x - 1)^n.$$

$$f'(x) = \sum_{n = 0}^{\infty} 2^n n(x - 1)^{n-1}.$$

3) Find $\int f(x) dx$ where

$$f(x) = \sum_{n = 1}^{\infty} \frac{(x - 3)^n}{n!}.$$

$$\int f(x) dx = C + \sum_{n = 0}^{\infty} \frac{1}{n!} \frac{(x - 3)^{n+1}}{n + 1}$$

$$= C + \sum_{n = 0}^{\infty} \frac{(x - 3)^{n+1}}{(n + 1)!}.$$

4) Is $f(x) = \sum_{n = 0}^{\infty} \frac{x^n}{4^n}$ continuous at $x = 2$?

Yes. 2 is in $[-4, 4]$, the interval of convergence for $f(x)$.

5) Give $\dfrac{1}{1-x} = \displaystyle\sum_{n=0}^{\infty} x^n$, $|x| < 1$, find a power series for $\dfrac{1}{(1+x)^2}$.

Substitute $-x$ for x:

$$\frac{1}{1+x} = \sum_{n=0}^{\infty} (-x)^n = \sum_{n=0}^{\infty} (-1)^n x^n.$$

Differentiate term by term:

$$\frac{-1}{(1+x)^2} = \sum_{n=0}^{\infty} (-1)^n n x^{n-1}.$$

Thus $\dfrac{1}{(1+x)^2} = (-1) \displaystyle\sum_{n=0}^{\infty} (-1)^n n x^{n-1}.$

$$= \sum_{n=0}^{\infty} (-1)^{n+1} n x^{n-1}.$$

Since $|-x| < 1$ is the same as $|x| < 1$, the radius of convergence is $R = 1$. At $x = 1$ we

have $\displaystyle\sum_{n=0}^{\infty} (-1)^{n+1} n$ and at $x = -1$ we have

$$\sum_{n=0}^{\infty} (-1)^{n+1} n (-1)^{n-1} = \sum_{n=0}^{\infty} n.$$

Both series diverge so the interval of convergence is $(-1, 1)$.

6) Find $\displaystyle\int \frac{1}{1+x^5}\, dx$ as a power series.

From $\dfrac{1}{1-x} = \displaystyle\sum_{n=0}^{\infty} x^n,$

$$\frac{1}{1+x} = \sum_{n=0}^{\infty} (-x)^n = \sum_{n=0}^{\infty} (-1)^n x^n.$$

Thus $\dfrac{1}{1+x^5} = \displaystyle\sum_{n=0}^{\infty} (-1)^n x^{5n}.$

$$\int \frac{1}{1+x^5}\, dx = \sum_{n=0}^{\infty} \int (-1)^n x^{5n}\, dx$$

$$= \sum_{n=0}^{\infty} \frac{(-1)^n}{5n+1} x^{5n+1} + C.$$

7) From $\dfrac{1}{1-x} = \displaystyle\sum_{n=0}^{\infty} x^n$, find the sum of the series $\displaystyle\sum_{n=1}^{\infty} n\left(\frac{1}{3}\right)^{n-1}.$

By differentiation, $\left(\dfrac{1}{1-x}\right)' = \displaystyle\sum_{n=1}^{\infty} n x^{n-1}.$

Thus $\displaystyle\sum_{n=1}^{\infty} n x^{n-1} = \frac{1}{(1-x)^2}.$

At $x = \dfrac{1}{3}$, $\dfrac{1}{\left(1 - \frac{1}{3}\right)^2} = \dfrac{9}{4}.$

Section 12.10   Taylor and Maclaurin Series

This section describes which functions may be written as power series and how to find such representations. The second partial sum of this representation will turn out to be our old friend—the equation of the tangent line. Partial sums of the series for a function will be useful for approximating values of the function.

Concepts to Master

A. Taylor series; Maclaurin series; Analytic functions

B. Taylor polynomials of degree n

C. Remainder of a Taylor series; Taylor's Inequality

D. Multiplication and division of power series

Summary and Focus Questions

Page 798 (ET Page 762)

A. If a function $y = f(x)$ can be expressed as a power series in $(x - a)$, that series must have the form:

$$f(x) = \sum_{n=0}^{\infty} \frac{f^{(n)}(a)}{n!}(x-a)^n$$

$$= f(a) + \frac{f'(a)}{1!}(x-a) + \frac{f''(a)}{2!}(x-a)^2 + \frac{f^3(a)}{3!}(x-a)^3 + \dots$$

This is called the *Taylor series for f at a*. In the special case of $a = 0$, it is called the *Maclaurin series for f*.

Not all functions can be represented by their power series; if f can be expressed as a power series in $(x - a)$, we say that f is *analytic about a*. If f is analytic about a, then f is equal to its Taylor series at a.

Here are some basic Taylor series about zero (Maclaurin series):

$$e^x = \sum_{n=0}^{\infty} \frac{x^n}{n!} \text{ for all } x. \qquad \ln(1+x) = \sum_{n=0}^{\infty} \frac{(-1)^{n+1}}{n} x^n \text{ for } x \in (-1, 1].$$

$$\sin x = \sum_{n=0}^{\infty} \frac{(-1)^n}{(2n+1)!} x^{2n+1} \text{ for all } x. \qquad \cos x = \sum_{n=0}^{\infty} \frac{(-1)^n}{(2n)!} x^{2n} \text{ for all } x.$$

$$\frac{1}{1-x} = \sum_{n=0}^{\infty} x^n \text{ for } |x| < 1. \qquad \tan^{-1} x = \sum_{n=0}^{\infty} (-1)^n \frac{x^{2n+1}}{2n+1} \text{ for } |x| \le 1.$$

To find a Taylor series for a given $y = f(x)$, you first find a formula for $f^{(n)}(a)$, usually in terms of n. Computing the first few derivatives $f'(a)$, $f''(a), f'''(a), f^{(4)}(a), \ldots$ often helps to see what the general term $f^{(n)}(a)$ looks like.

For some functions f it is easier to find the Taylor series for $f'(x)$ or $\int f(x)dx$ then integrate or differentiate term by term to obtain the series for f. In some other cases, substitutions in the basic Taylor series and algebra can be used to find the Taylor series for f.

1) Find the Taylor series for $f(x) = \frac{1}{x}$ at $a = 1$.

    a) directly from the definition.

First find the general form of $f^{(n)}(1)$:

$$f(x) = x^{-1} \qquad\qquad f(1) = 1$$
$$f'(x) = -x^{-2} \qquad\quad f'(1) = -1$$
$$f''(x) = 2x^{-3} \qquad\quad f''(1) = 2$$
$$f'''(x) = -6x^{-4} \qquad f'''(1) = -6$$
$$f^{(4)}(x) = 24x^{-5} \qquad f^{(4)}(1) = 24$$

In general, $f^{(n)}(1) = (-1)^n n!$.
Thus the Taylor series is

$$\sum_{n=0}^{\infty} \frac{(-1)^n n!}{n!} (x-1)^n = \sum_{n=0}^{\infty} (-1)^n (x-1)^n$$

$$= \sum_{n=0}^{\infty} (1-x)^n.$$

    b) using substitution in a geometric series

$$\left(\text{Hint: } \frac{1}{x} = \frac{1}{1-(1-x)}\right).$$

In the form $\frac{1}{1-(1-x)}$, this is the value of a geometric series with $a = 1, r = 1 - x$.

Thus $\dfrac{1}{x} = \displaystyle\sum_{n=1}^{\infty} 1(1-x)^{n-1} = \sum_{n=0}^{\infty} (1-x)^n.$

    c) What is the interval of convergence for your answer to part a)?

Apply the Ratio Test:

$$\lim_{n\to\infty} \left| \frac{(1-x)^{n+1}}{(1-x)^n} \right| = \lim_{n\to\infty} |1 - x| = |1 - x|.$$

$|1 - x| < 1$ is equivalent to $0 < x < 2$.
At $x = 0$ and $x = 2$ the terms of

$$\sum_{n=0}^{\infty} (1-x)^n \text{ do no approach zero so those}$$

series diverge. The interval of convergence is $(0, 2)$.

2) Find directly the Maclaurin series for
$f(x) = e^{4x}$.

$$f(x) = e^{4x} \qquad f(0) = 1$$
$$f'(x) = 4e^{4x} \qquad f'(0) = 4$$
$$f''(x) = 16e^{4x} \qquad f''(0) = 16$$

In general $f^{(n)}(0) = 4^n$.

Thus $e^{4x} = \sum_{n=0}^{\infty} \frac{4^n}{n!}x^n$.

3) True, False:
If f is analytic at a, then

$$f(x) = \sum_{n=0}^{\infty} \frac{f^{(n)}(a)}{n!} (x - a)^n.$$

True.

4) Obtain the Maclaurin series for $\frac{x}{1 + x^2}$ from
the series for $\frac{1}{1 - x} = \sum_{n=0}^{\infty} x^n$.

$$\frac{1}{1 - x} = 1 + x + x^2 + \dots x^n + \dots$$

Substitute $-x^2$ for x:

$$\frac{1}{1 + x^2} = 1 - x^2 + x^4 - x^6 + \dots$$
$$+ (-1)^n x^{2n} + \dots$$

Now multiply by x:

$$\frac{x}{1 + x^2} = x - x^3 + x^5 - x^7 + \dots$$
$$+ (-1)^n x^{2n+1} + \dots$$
$$= \sum_{n=0}^{\infty} (-1)^n x^{2n+1}.$$

5) Using $e^x = \sum_{n=0}^{\infty} \frac{x^n}{n!}$ find the Maclaurin series
for $\sinh x$.
$\left(\text{Remember, } \sinh x = \frac{e^x - e^{-x}}{2}.\right)$

$$e^x = 1 + x + \frac{x^2}{2!} + \dots \frac{x^n}{n!} + \dots$$

Thus $e^{-x} = 1 - x + \frac{x^2}{2!} + \dots \frac{(-1)^n x^n}{n!} + \dots$

Subtracting term by term (the even terms cancel)

$$e^x - e^{-x} = 2x + \frac{2x^3}{3!} + \dots + \frac{2x^{2n+1}}{(2n + 1)!} + \dots$$

Thus $\sinh x = x + \frac{x^3}{3!} + \dots + \frac{x^{2n+1}}{(2n + 1)!} + \dots$

$$= \sum_{n=0}^{\infty} \frac{x^{2n+1}}{(2n + 1)!}.$$

6) Find the infinite series expression for
$$\int_0^{1/2} \frac{1}{1 - x^3}\, dx.$$

$$\frac{1}{1 - x} = \sum_{n=0}^{\infty} x^n. \text{ Thus } \frac{1}{1 - x^3} = \sum_{n=0}^{\infty} x^{3n}.$$

$$\int \frac{1}{1 - x^3}\, dx = \int \sum_{n=0}^{\infty} x^{3n}\, dx$$

$$= \sum_{n=0}^{\infty} \frac{x^{3n+1}}{3n + 1}.$$

$$\int_0^{1/2} \frac{1}{1 - x^3}\, dx = \sum_{n=0}^{\infty} \frac{x^{3n+1}}{3n + 1}\Bigg]_0^{1/2}$$

$$= \sum_{n=0}^{\infty} \frac{\left(\frac{1}{2}\right)^{3n+1}}{3n + 1} - 0 = \sum_{n=0}^{\infty} \frac{1}{(6n + 2)8^n}.$$

Page 799 (ET Page 763)

B. If $f^{(n)}(a)$ exists, the *Taylor polynomial of degree n for f about a* is the nth partial sum of the Taylor series:

$$T_n(x) = f(a) + \frac{f'(a)}{1!}(x - a) + \frac{f''(a)}{2!}(x - a)^2 + \frac{f^{(3)}(a)}{3!}(x - a)^3 + \dots \frac{f^{(n)}(a)}{n!}(x - a)^n.$$

The Taylor polynomial of degree one, $T_1(x) = f(a) + f'(a)(x - a)$, is the familiar equation for the tangent line to $y = f(x)$ at a point a.

$T_2(x) = f(a) + f'(a)(x - a) + \frac{1}{2}f''(a)(x - a)^2$ is the "tangent parabola" to $y = f(x)$.

Since the first n derivatives of $T_n(x)$ at a are equal to the corresponding first n derivatives of $f(x)$ at a, $T_n(x)$ is an approximation for $f(x)$ if x is near a.

7) Let $f(x) = \sqrt{x}$.

a) Construct the Taylor polynomial of degree 3 for $f(x)$ about $x = 1$.

$$f(x) = x^{1/2} \qquad\qquad f(1) = 1$$
$$f'(x) = \tfrac{1}{2}x^{-1/2} \qquad\qquad f'(1) = \tfrac{1}{2}$$
$$f''(x) = -\tfrac{1}{4}x^{-3/2} \qquad\qquad f''(1) = -\tfrac{1}{4}$$
$$f^{(3)}(x) = \tfrac{3}{8}x^{-5/2} \qquad\qquad f^{(3)}(1) = \tfrac{3}{8}$$

Therefore, $T_3(x)$

$$= 1 + \tfrac{1}{2}(x - 1) - \frac{1/4}{2!}(x - 1)^2 + \frac{3/8}{3!}(x - 1)^3$$
$$= 1 + \tfrac{1}{2}(x - 1) - \tfrac{1}{8}(x - 1)^2 + \tfrac{1}{16}(x - 1)^3$$

b) Approximate $\sqrt{\tfrac{3}{2}}$ using the Taylor polynomial of degree 3 from part a).

$$T_3\left(\tfrac{3}{2}\right) = 1 + \tfrac{1}{2}\left(\tfrac{1}{2}\right) - \tfrac{1}{8}\left(\tfrac{1}{2}\right)^2 + \tfrac{1}{16}\left(\tfrac{1}{2}\right)^3 = \frac{157}{128}$$
$$\approx 1.2265.$$

(To 4 decimals, $\sqrt{\tfrac{3}{2}} = 1.2247$.)

8) a) Find the Taylor polynomial of degree 6 for $f(x) = \cos x$ about $x = 0$.

$$f(0) = \cos 0 = 1$$
$$f'(x) = -\sin x \qquad f'(0) = 0$$
$$f''(x) = -\cos x \qquad f''(0) = -1$$
$$f^{(3)}(x) = \sin x \qquad f^{(3)}(0) = 0$$
$$f^{(4)}(x) = \cos x \qquad f^{(4)}(0) = 1$$
$$f^{(5)}(0) = 0 \text{ and} \qquad f^{(6)}(0) = -1.$$

Thus $T_6(x) = 1 + \frac{-1}{2!}x^2 + \frac{1}{4!}x^4 - \frac{1}{6!}x^6$

$$= 1 - \frac{x^2}{2} + \frac{x^4}{24} - \frac{x^6}{720}.$$

Note, of course, that this is the first few terms of the Maclaurin series for cosine.

b) Use your answer to part a) to approximate cos 1.

$$\cos 1 \approx T_6(1) = 1 - \frac{(1)^2}{2} + \frac{(1)^4}{24} - \frac{(1)^6}{720}$$

$$= 1 - 0.5 + 0.041667 - 0.001389$$

$$= 0.540278$$

(Note: cos 1 = 0.540302 to 6 decimal places.)

Page 799 (ET Page 763)

C. If $f^{(n)}(x)$ exists for $|x - a| < d$, the nth *remainder of the Taylor series of degree n for f* is:

$$R_n(x) = f(x) - T_n(x)$$

(The R used in $R_n(x)$ should not be confused with the R used earlier in the chapter for the radius of convergence of an infinite series.)

Taylor's Inequality: If $|f^{(n+1)}(x)| \le M$ for $|x - a| \le d$, then the nth remainder $R_n(x)$ of the Taylor series satisfies the inequality

$$|R_n(x)| \le \frac{M}{(n+1)!}|x - a|^{n+1} \text{ for } |x - a| \le d.$$

If f has derivatives of all orders and $\lim\limits_{n \to \infty} R_n(x) = 0$ for $|x - a| < d$, then $f(x)$ is equal to its Taylor series for $|x - a| \le d$. This is one way to show that a function is equal to its Taylor series—show $\lim\limits_{n \to \infty} R_n(x) = 0$. You will often need to use this limit:

$$\lim_{n \to \infty} \frac{x^n}{n!} = 0 \text{ for all real numbers } x.$$

9) True or False:
In general, the larger the number n, the closer the Taylor polynomial approximation is to the actual functional value.

True.

10) a) Find an expression for the nth Taylor polynomial $T_n(x)$ for $f(x) = 2^x$ at $x = 0$.

$$f(x) = 2^x \qquad\qquad f(0) = 1$$
$$f'(x) = (\ln 2)\, 2^x \qquad f'(0) = \ln 2$$
$$f''(x) = (\ln 2)^2\, 2^x \qquad f''(0) = (\ln 2)^2$$
and, in general,
$$f^n(x) = (\ln 2)^n\, 2^x \qquad f^n(0) = (\ln 2)^n.$$
$$T_n(x) = 1 + (\ln 2)x + \frac{(\ln 2)^2}{2!}x^2 + \dots$$
$$\qquad + \frac{(\ln 2)^n}{n!}\, x^n.$$

b) Find an upper bound for $\left|f^{(n+1)}(x)\right|$ for $|x| \le 1$.

$$\left|f^{(n+1)}(x)\right| = \left|(\ln 2)^{n+1} \cdot 2^x\right|$$
$$= (\ln 2)^{n+1} \cdot 2^x \le (\ln 2)^{n+1} \cdot 2 \text{ since}$$
$$|x| \le 1.$$

c) Show that $f(x) = 2^x$ is equal to its Taylor series for $|x| < 1$.

We need to show that $\lim_{n\to\infty} R_n(x) = 0$ for $|x| \le 1$. From part b)
$$\left|f^{(n+1)}(x)\right| \le 2\,(\ln 2)^{n+1}.$$
By Taylor's Inequality
$$0 \le |R_n(x)| \le \frac{2(\ln 2)^{n+1}}{(n+1)!}|x|^{n+1} \le \frac{2(\ln 2)^{n+1}}{(n+1)!}$$
since $|x| \le 1$.
$$\lim_{n\to\infty} 2\frac{(\ln 2)^{n+1}}{(n+1)!} = 2\lim_{n\to\infty}\frac{(\ln 2)^{n+1}}{(n+1)!} = 2(0) = 0.$$
Therefore, by the Squeeze Theorem for limits at infinity, $\lim_{n\to\infty} R_n(x) = 0$.

11) Find $\int_0^1 xe^{-x}\,dx$ using a series to within 0.01.

$$e^x = \sum_{n=0}^{\infty} \frac{x^n}{n!}, \quad e^{-x} = \sum_{n=0}^{\infty} \frac{(-1)^n x^n}{n!},$$

so $xe^{-x} = \sum_{n=0}^{\infty} \frac{(-1)^n x^{n+1}}{n!}$. Thus

$$\int_0^1 xe^{-x}\,dx = \int_0^1 \sum_{n=0}^{\infty} \frac{(-1)^n x^{n+1}}{n!}\,dx$$

$$= \sum_{n=0}^{\infty} \frac{(-1)^n x^{n+2}}{n!(n+2)}\Big]_0^1 = \sum_{n=0}^{\infty} \frac{(-1)^n}{n!(n+2)} - 0$$

$$= \sum_{n=0}^{\infty} \frac{(-1)^n}{n!(n+2)}$$

$$= \frac{1}{0!2} - \frac{1}{1!3} + \frac{1}{2!4} - \frac{1}{3!5} + \frac{1}{4!6} - \frac{1}{5!7} + \cdots$$

$$= \frac{1}{2} - \frac{1}{3} + \frac{1}{8} - \frac{1}{30} + \frac{1}{144} + \cdots$$

This is an alternating series and the fifth term $\frac{1}{144}$ is less than 0.01. Thus,

$$\int_0^1 xe^{-x}\,dx \approx \frac{1}{2} - \frac{1}{3} + \frac{1}{8} - \frac{1}{30} \approx 0.2583.$$

Using integration by parts:

$$\int_0^1 xe^{-x}\,dx = (-xe^{-x} - e^{-x})\Big]_0^1 = 1 - \frac{2}{e}.$$

Thus our estimate is within 0.01 of the actual value $1 - \frac{2}{e} \approx 0.2642$.

**Page 805
(ET Page 769)**

D. Convergent series may be multiplied and divided like polynomials. Often finding the first few terms of the result is sufficient.

12) Find the first three terms of the Maclaurin series for $e^x \sin x$.

$$e^x = 1 + x + \frac{x^2}{2} + \frac{x^3}{6} + \cdots$$

$$\sin x = x - \frac{x^3}{3!} + \frac{x^5}{5!} - \cdots$$

Thus

$$e^x \sin x = x\left(1 + x + \frac{x^2}{2} + \frac{x^3}{6} + \cdots\right)$$

$$- \frac{x^3}{6}\left(1 + x + \frac{x^2}{2} + \frac{x^3}{6} + \cdots\right)$$

$$+ \frac{x^5}{120}\left(1 + x + \frac{x^2}{2} + \frac{x^3}{6} + \cdots\right) + \cdots$$

$$= \left(x + x^2 + \frac{x^3}{2} + \frac{x^4}{6} + \cdots\right)$$

$$- \left(\frac{x^3}{6} + \frac{x^4}{6} + \frac{x^5}{12} + \frac{x^6}{36} + \cdots\right)$$

$$+ \left(\frac{x^5}{120} + \frac{x^6}{120} + \frac{x^7}{240} + \cdots\right)$$

$$= x + x^2 + \frac{x^3}{6} + \cdots$$

Section 12.11 The Binomial Series

You learned in your algebra class that the Binomial Theorem may be used to expand $(1 + x)^k$:

$$(1 + x)^k = 1^k + \binom{k}{1}1^{k-1}x + \binom{k}{2}1^{k-2}x^2 + \binom{k}{3}1^{k-3}x^3 + \ldots + \binom{k}{k-1}1x^{k-1} + x^k$$

$$= 1 + \binom{k}{1}x + \binom{k}{2}x^2 + \binom{k}{3}x^3 + \ldots + \binom{k}{k-1}x^{k-1} + x^k$$

where $\binom{k}{r} = \frac{k!}{r!(k-r)!} = \frac{k(k-1)\ldots(k-r+1)}{r!}$ is a binomial coefficient. This section extends the expansion of $(1 + x)^k$ to the case where k is no longer a positive integer. In the general case, the expansion of $(1 + x)^k$ is an infinite series. The Binomial Theorem becomes a special case in which only the first few terms of the infinite series are non-zero.

Concepts to Master

Binomial coefficient; Binomial series for $(1 + x)^k$

Page 808 (ET Page 772)

Summary and Focus Questions

For any real number k and a non-negative integer n, the *binomial coefficient*

$$\binom{k}{n} = \begin{cases} 1 & \text{if } n = 0 \\ \dfrac{k(k-1)(k-2)\ldots(k-n+1)}{n!} & \text{if } n \geq 1 \end{cases}$$

When k is a positive integer, this is the binomial coefficient that you have seen in an algebra class.

For any real number k the *binomial series for* $(1 + x)^k$ is the Maclaurin series for $(1 + x)^k$:

$$(1 + x)^k = \sum_{n=0}^{\infty} \binom{k}{n}x^n = 1 + kx + \frac{k(k-1)}{2!}x^2 + \ldots + \binom{k}{n}x^n + \ldots$$

The binomial series converges to $(1 + x)^k$ for these cases:

Condition on k	Interval of Convergence
$k \leq -1$	$(-1, 1)$
$-1 < k < 0$	$(-1, 1]$
k nonnegative integer	$(-\infty, \infty)$
$k > 0$, but not an integer	$[-1, 1]$

1) a) Find $\binom{6}{4}$.

$$\binom{6}{4} = \frac{6(5)(4)(3)}{4!} = 15$$

b) Find $\binom{4/3}{5}$.

$$\binom{4/3}{5} = \frac{\left(\frac{4}{3}\right)\left(\frac{1}{3}\right)\left(-\frac{2}{3}\right)\left(-\frac{5}{3}\right)\left(-\frac{8}{3}\right)}{5 \cdot 4 \cdot 3 \cdot 2 \cdot 1} = \frac{-8}{3^6} = -\frac{8}{729}.$$

2) True or False:
If k is a nonnegative integer, then the binomial series for $(1 + x)^k$ is a finite sum.

True, only the first $k + 1$ terms of the infinite series are non-zero.

3) Find the binomial series for $\sqrt[3]{1 + x}$.

$$\sum_{n=0}^{\infty} \binom{\frac{1}{3}}{n} x^n.$$

4) Find a power series for $f(x) = \sqrt{4 + x}$.

$$\sqrt{4 + x} = 2\left(1 + \frac{x}{4}\right)^{1/2} = 2\sum_{n=0}^{\infty} \binom{\frac{1}{2}}{n}\left(\frac{x}{4}\right)^n$$
$$= 2\sum_{n=0}^{\infty} \binom{\frac{1}{2}}{n}\frac{x^n}{4^n}.$$

5) Find an infinite series for $\sin^{-1} x$. (Hint: Start with $(1 + x)^{-1/2}$.)

$$\frac{1}{\sqrt{1 + x}} = (1 + x)^{-1/2} = \sum_{n=0}^{\infty} \binom{-\frac{1}{2}}{n}x^n.$$

Substitute $-x$ for x:

$$\frac{1}{\sqrt{1 - x}} = \sum_{n=0}^{\infty} \binom{-\frac{1}{2}}{n}(-x)^n$$
$$= \sum_{n=0}^{\infty} (-1)^n \binom{-\frac{1}{2}}{n}x^n.$$

Substitute x^2 for x:

$$\frac{1}{\sqrt{1 - x^2}} = \sum_{n=0}^{\infty} (-1)^n \binom{-\frac{1}{2}}{n}x^{2n}.$$

Now integrate term by term:

$$\sin^{-1} x = \sum_{n=0}^{\infty} \frac{(-1)^n \binom{-\frac{1}{2}}{n}x^{2n+1}}{2n + 1}.$$

Section 12.12   Applications of Taylor Polynomials

This last section uses Taylor polynomials to estimate values of functions and uses Taylor's Inequality to measure the accuracy of such approximations.

Concepts to Master

A. Approximate functional values using a Taylor polynomial
B. Estimate the error in Taylor polynomial approximations

Summary and Focus Questions

Page 813 (ET Page 777)

A. The Taylor polynomial of degree n for f about a

$$T_n(x) = f(a) + \frac{f'(a)}{1!}(x - a) + \frac{f''(a)}{2!}(x - a)^2 + \frac{f^{(3)}(a)}{3!}(x - a)^3 + \dots \frac{f^{(n)}(a)}{n!}(x - a)^n$$

may be used to approximate $f(x)$ for any given x in the interval of convergence for the Taylor series for f about a.

1) a) For $f(x) = \frac{1}{3 + x}$, find $T_3(x)$ where $a = 2$.

$$f(x) = (3 + x)^{-1} \qquad f(2) = \frac{1}{5}$$
$$f'(x) = -(3 + x)^{-2} \qquad f'(2) = \frac{-1}{25}$$
$$f''(x) = 2(3 + x)^{-3} \qquad f''(2) = \frac{2}{125}$$
$$f'''(x) = -6(3 + x)^{-4} \qquad f'''(2) = \frac{-6}{625}$$

$$T_3(x) = \frac{1}{5} - \frac{1}{25}(x - 2) + \frac{2/125}{2!}(x - 2)^2$$
$$- \frac{6/625}{3!}(x - 2)^3$$
$$= \frac{1}{5} - \frac{1}{25}(x - 2) + \frac{1}{125}(x - 2)^2$$
$$- \frac{1}{625}(x - 2)^3.$$

b) Estimate $f(3)$ with $T_3(3)$.

$$T_3(3) = \frac{1}{5} - \frac{(3 - 2)}{25} + \frac{(3 - 2)^2}{125} - \frac{(3 - 2)^3}{625}$$
$$= \frac{1}{5} - \frac{1}{25} + \frac{1}{125} - \frac{1}{625}$$
$$= \frac{104}{625} \approx 0.1664$$

We note that this is a lot of work to approximate $f(3) = \frac{1}{6}$.

2) a) For $f(x) = x^{5/2}$, find $T_3(x)$ where $a = 4$.

$$f(x) = x^{5/2} \qquad\qquad f(4) = 32$$
$$f'(x) = \tfrac{5}{2}x^{3/2} \qquad\quad f'(4) = 20$$
$$f''(x) = \tfrac{15}{4}x^{1/2} \qquad\; f''(4) = \tfrac{15}{2}$$
$$f'''(x) = \tfrac{15}{8}x^{-1/2} \qquad f'''(4) = \tfrac{15}{16}$$

$$T_3(x) = 32 + 20(x - 4) + \tfrac{15/2}{2!}(x - 4)^2$$
$$\qquad\qquad + \tfrac{15/16}{3!}(x - 4)^3$$
$$= 32 + 20(x - 4) + \tfrac{15}{4}(x - 4)^2$$
$$\qquad\qquad + \tfrac{5}{32}(x - 4)^3.$$

b) Estimate $(5)^{5/2}$ with $T_3(x)$.

$$T_3(5) = 32 + 20(1)^1 + \tfrac{15}{4}(1)^2 + \tfrac{5}{32}(1)^3$$
$$= \tfrac{1789}{32} \approx 55.9063.$$

To four decimals $(5)^{5/2}$ is approximately 55.9017.

Page
813
(ET Page
777)

B. The accuracy of the approximation $f(x) \approx T_n(x)$ will usually depend on n, x, and a. In general:

n — the larger the value of n, the better the estimate

x and a — the closer that x is to a, the better the estimate.

From section 11.10, we may use Taylor's Inequality:

If $\left|f^{(n+1)}(x)\right| \le M$ for $|x - a| \le d$, then $\left|R_n(x)\right| \le \dfrac{M}{(n+1)!}|x - a|^{n+1}$ for $|x - a| \le d$,

to estimate the error. The approximation $T_n(x)$ will be accurate to within any given number ϵ by selecting n such that

$$\frac{M}{(n+1)!}|x - a|^{n+1} < \epsilon.$$

In the special case where the Taylor series is an alternating series, $\left|R_n(x)\right|$ may be estimated with $\left|\dfrac{f^{(n+1)}(x)}{(n+1)!}(x - a)^{n+1}\right|$, the $(n+1)^{\text{st}}$ term of the Taylor series for f.

3) a) Use Taylor's Inequality to estimate the accuracy of $f(x) \approx T_3(x)$ for $1 < x < 4$ for question 1).

From 1 a), $f^{(4)}(x) = 24(3 + x)^{-5}$.
If $1 \le x \le 4$, then $4 \le x + 3 \le 7$.
Therefore $4^5 \le (x + 3)^5 \le 7^5$, and

$$\frac{1}{7^5} \le (x + 3)^{-5} \le \frac{1}{4^5} \text{ and}$$

$$\frac{24}{7^5} \le 24(x + 3)^{-5} \le \frac{24}{4^5}$$

Thus $\left| f^{(4)}(x) \right| \le \frac{24}{4^5}$.

From $1 \le x \le 4$, $-1 \le x - 2 \le 2$.
Thus $(x - 2)^4 \le 2^4 = 16$.
Finally,

$$\left| R_3(x) \right| \le \frac{\frac{24}{4^5}}{4!} (x - 2)^4$$

$$\le \frac{24}{4^5 \cdot 4!} \cdot 16 = 0.0156.$$

b) Check the accuracy of the estimate when $x = 3$.

$f(3) = \frac{1}{3 + 3} = \frac{1}{6} \approx 0.1667$.
$T_3(3) \approx 0.1664$ by 1 b).
Thus $R_3(3) \approx 0.1667 - 0.1664 = 0.0003$,
which is within 0.0156.

4) a) For $f(x) = x^{5/2}$, estimate the error between $f(5)$ and $T_3(5)$. (See question 2.)

From 2 a), $f^{(4)}(x) = -\frac{15}{16}x^{-3/2}$.

For $|x - 4| \le 2$, $2 \le x \le 6$.

$6^{-3/2} \le x^{-3/2} \le 2^{-3/2}$.

Thus $\left| f^{(4)}(x) \right| \le \frac{15}{16} \cdot 2^{-3/2} = \frac{15}{32\sqrt{2}}$.

Therefore $\left| R_3(x) \right| \le \frac{15}{32\sqrt{2}}|x - 4|^4$

$$\le \frac{\frac{15}{32\sqrt{2}}}{4!}(2)^4 \approx 0.221.$$

b) Check the accuracy of the estimate $f(5) \approx T_3(5)$.

$f(5) = 5^{5/2} \approx 55.9017$.
$T_3(5) \approx 55.9063$ (question 2b)).
$R_3(5) \approx 55.9017 - 55.9063 = -0.0046$,
which is within 0.221.

5) a) What degree Maclaurin polynomial is needed to approximate cos 0.5 accurate to within 0.00001?

Rephrased, the question asks: For what n is $|R_n(0.5)| < 0.00001$ when $f(x) = \cos x$ and $a = 0$?

Because $f^{(n)}(x) = \pm\sin x$ or $\pm\cos x$ for all n, $|f^{(n+1)}(x)| \le 1$.

Thus $|R_n(x)| \le \frac{1}{(n+1)!}|x|^{n+1}$

$= \frac{|x|^{n+1}}{(n+1)!} \le \frac{1}{(n+1)!}$ if $|x| \le 1$.

$\frac{1}{(n+1)!} \le 0.00001$ means

$(n+1)! \ge 100{,}000$.

Therefore $n = 8$, since $8! = 40{,}320$ and $9! = 362{,}800$.

b) Use your answer to part a) to estimate cos 0.5 to within 0.00001.

$T_8(x) = 1 - \frac{x^2}{2} + \frac{x^4}{24} - \frac{x^6}{720} + \frac{x^8}{40320}$

$T_8(0.5) = 1 - \frac{(0.5)^2}{2} + \frac{(0.5)^4}{24} - \frac{(0.5)^6}{720} +$

$\frac{(0.5)^8}{40320} \approx 0.877583.$

To 6 decimals, cos 0.5 = 0.877583 so the approximation is within 0.00001.

Technology Plus for Chapter 12

1) Using a spreadsheet or calculator, find the first 20 partial sums of $\displaystyle\sum_{n=1}^{\infty} \frac{1}{n^4 + 1}$. Estimate the sum of the series.

n	$\dfrac{1}{n^4 + 1}$	$\displaystyle\sum_{k=1}^{n} \frac{1}{k^4 + 1}$
1	0.50000	0.50000
2	0.05882	0.55882
3	0.01220	0.57102
4	0.00389	0.57491
5	0.00160	0.57651
6	0.00077	0.57728
7	0.00042	0.57769
8	0.00024	0.57794
9	0.00015	0.57709
10	0.00010	0.57819
11	0.00007	0.57826
12	0.00005	0.57831
13	0.00004	0.57834
14	0.00003	0.57837
15	0.00002	0.57839
16	0.00002	0.57840
17	0.00001	0.57842
18	0.00001	0.57843
19	0.00001	0.57843

$$\sum_{n=1}^{\infty} \frac{1}{n^4 + 1} \approx 0.57844.$$

2) Use the 15th partial sum to estimate
$$\sum_{n=1}^{\infty} \frac{(-1)^{n-1}}{n^3}.$$ How accurate is the estimate?

n	$\frac{(-1)^{n-1}}{n^3}$	$\sum_{k=1}^{n} \frac{(-1)^{k-1}}{k^3}$
1	1.00000	1.00000
2	−0.12500	0.87500
3	0.03704	0.91204
4	−0.01563	0.89641
5	0.00800	0.90441
6	−0.00463	0.89978
7	0.00292	0.90270
8	−0.00195	0.90074
9	0.00137	0.90212
10	−0.00100	0.90112
11	0.00075	0.90187
12	−0.00058	0.90129
13	0.00046	0.90174
14	−0.00036	0.90138
15	0.00030	0.90168

The 15th partial sum is

$$\sum_{n=1}^{15} \frac{(-1)^{n-1}}{n^3} \approx 0.90168.$$

This is within $\left| \frac{(-1)^{n-1}}{16^3} \right| = 0.00024$

of the sum of the series.

3) On the same screen sketch a graph of
$f(x) \dfrac{1}{\sqrt{x}}$ and its first 3 Taylor polynomials
about $x = 1$.

$f(x) = x^{-1/2}$, $f(1) = 1$

$f'(x) = -\frac{1}{2}x^{-3/2}$, $f'(1) = -\frac{1}{2}$

$f''(x) = \frac{3}{4}x^{-5/2}$, $f''(1) = \frac{3}{4}$

$f'''(x) = -\frac{15}{8}x^{-7/2}$, $f'''(1) = -\frac{15}{8}$

$TP_1(x) = 1 + (-\frac{1}{2})(x - 1) = 1 - \frac{1}{2}(x - 1)$

$TP_2(x) = 1 - \frac{1}{2}(x - 1) + \dfrac{\frac{3}{4}}{2!}(x - 1)^2$

$= 1 - \frac{1}{2}(x - 1) + \frac{3}{8}(x - 1)^2$

$TP_3(x) = 1 - \frac{1}{2}(x - 1) + \frac{3}{8}(x - 1)^2$

$+ \dfrac{-\frac{15}{8}}{3!}(x - 1)^3$

$= 1 - \frac{1}{2}(x - 1) + \frac{3}{8}(x - 1)^2 - \frac{15}{48}(x - 1)^3$

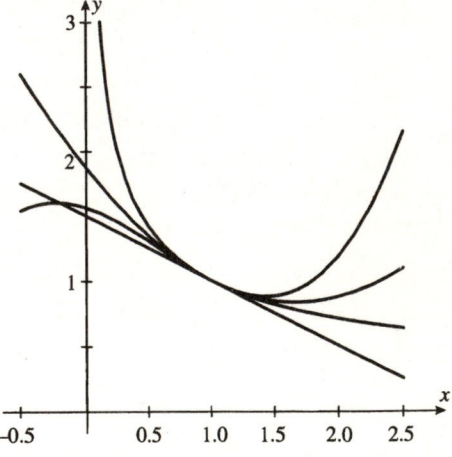

Chapter 13 — Vectors and the Geometry of Space

Cartoons courtesy of Sidney Harris. Used by permission.

Section 13.1   Three-Dimensional Coordinate Systems

This section extends the two dimensional *x*-, *y*- coordinate system to three dimensions (x, y, z) with three axes, each perpendicular to the other axes which you might visualize like the corner of a room.

Concepts to Master

A. Rectangular three-dimensional coordinates; Distance between two points

B. Planes, spheres, and regions in space

Summary and Focus Questions

**Page 829
(ET Page 793)***

A. With the *x*, *y*, and *z* axes perpendicular to each other in three dimensional space, each triple (a, b, c) of real numbers corresponds to a unique point in space.

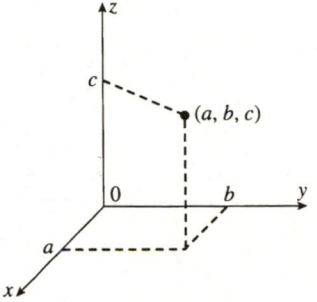

The distance between points P_1: (x_1, y_1, z_1) and P_2: (x_2, y_2, z_2) is

$$|P_1P_2| = \sqrt{(x_2 - x_1)^2 + (y_2 - y_1)^2 + (z_2 - z_1)^2}.$$

1) Plot the points *A*: $(0, 5, 0)$ *B*: $(5, 4, 6)$
   C: $(1, -1, 3)$.

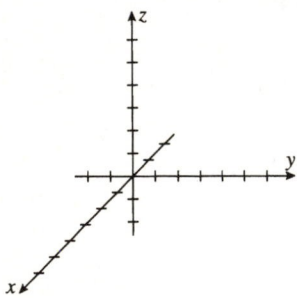

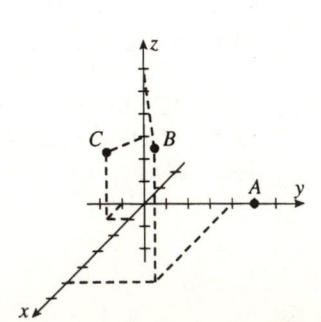

Remember, when using the Early Transcendentals book, use the page in parentheses!

2) The coordinates of the points A and B pictured are _____.

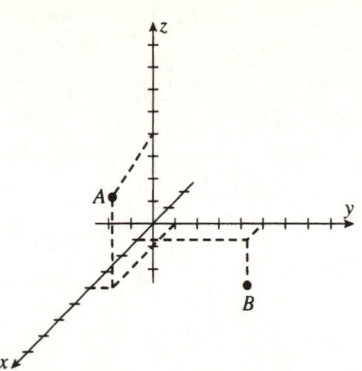

$A: (4, 1, 4)$ $B: (1, 5, -2)$

3) The distance between the points $(7, 5, -2)$ and $(4, 6, 1)$ is _____.

$$\sqrt{(7-4)^2+(5-6)^2+(-2-1)^2}=\sqrt{19}.$$

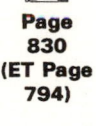

Page 830 (ET Page 794)

B. Some simple linear equations in x, y, z with one or more of the variables missing represent planes parallel to the axes of the missing variable. Only a portion of each plane is drawn here:

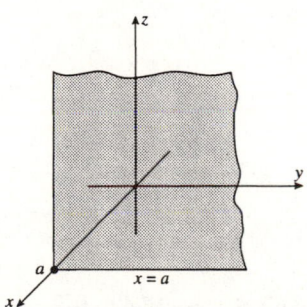

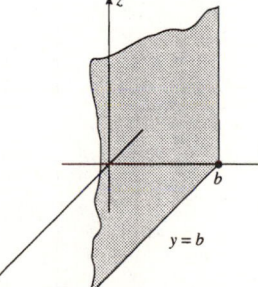

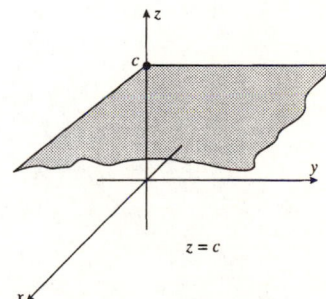

If a variable is missing, the graph will be parallel to the axis of the missing variable. For example, the graph if $3x + 2z = 6$ is a plane parallel to the y−axis.

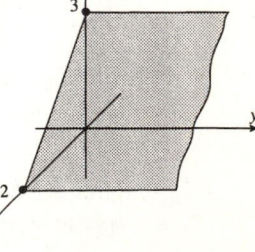

The equation of the sphere with center (a, b, c) and radius R is $(x - a)^2 + (y - b)^2 + (z - c)^2 = R^2$.

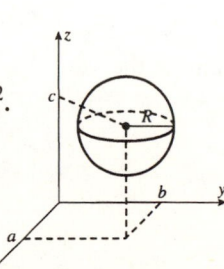

4) Describe the coordinates of each set of points:

a)

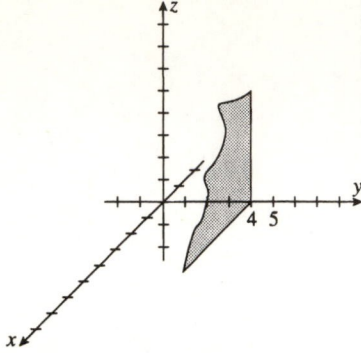

A point in the plane in the figure may have any values for x and z coordinates but must have a y coordinate of 4. The plane is all points (x, y, z) such that $y = 4$.

b)

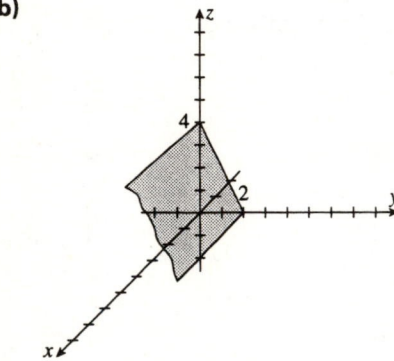

The plane is parallel to the x-axis so it has no x variable. In the yz-plane it is the line through $(0, 0, 4)$ and $(0, 2, 0)$: $z + 2y = 4$. The plane is $z + 2y = 4$.

5) Find the equation of the sphere with center $(2, 3, -4)$ and radius 5.

$(x - 2)^2 + (y - 3)^2 + (z + 4)^2 = 25.$

6) Find the center and radius of the sphere $x^2 - 6x + y^2 + 4y + z^2 = 1$.

Complete the square for x and y:
$$x^2 - 6x + 9 + y^2 + 4y + 4 + z^2$$
$$= 1 + 9 + 4$$
$$(x - 3)^2 + (y + 2)^2 + z^2 = 14$$
center: $(3, -2, 0)$
radius: $\sqrt{14}$.

7) Describe the region given by each:

a) $x^2 + z^2 = 4$

This is a cylinder of radius 2 along the y-axis.

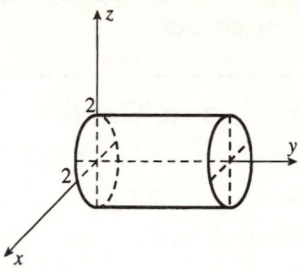

b) $x^2 + y^2 + z^2 \leq 1$

A solid ball of radius 1 and center $(0, 0, 0)$.

c) $\{(x, y, z): x = 3 \text{ and } y = 4\}$

This is a line through $(3, 4, 0)$ and parallel to the z-axis.

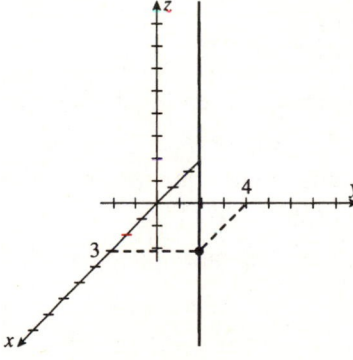

d) $|y| + |z| \leq 1$

This is a two by two square solid centered along the x-axis.

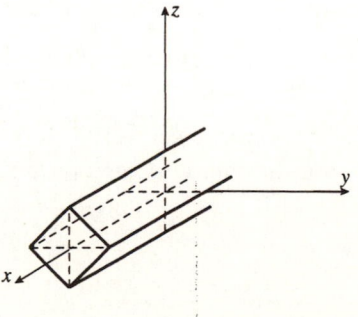

Section 13.2   Vectors

A vector has both a length and a direction, such as the force of wind at a given place and given time. This section studies two- and three-dimensional vectors and their properties.

Concepts to Master

A. Vectors; Position vector; Length; Unit Vector; Standard basis vectors (**i**, **j**, **k**)

B. Vector arithmetic; Scalars; Vector properties

Summary and Focus Questions

**Page 834
(ET Page 798)**

A. A *vector* is an object having both a magnitude and a direction. A *two-dimensional vector* is an ordered pair $\mathbf{a} = \langle a_1, a_2 \rangle$. A *three-dimensional vector* is an ordered triple $\mathbf{a} = \langle a_1, a_2, a_3 \rangle$. The numbers a_i are called *components* of **a**.

The directed line segment $\overrightarrow{AB}$ from $A: (x_1, y_1)$ to $B: (x_2, y_2)$ represents the vector
$\mathbf{a} = \langle x_2 - x_1, y_2 - y_1 \rangle$.

a is the *position vector* for the line segment from the origin O to $P: (x_2 - x_1, y_2 - y_1)$.

Both $\overrightarrow{AB}$ and $\overrightarrow{OP}$ represent the same vector.

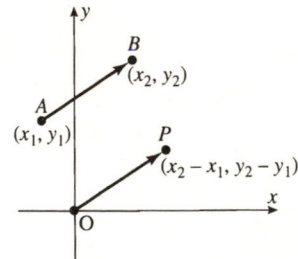

For $\mathbf{a} = \langle a_1, a_2 \rangle$, the *length (or magnitude) of* **a** is
$|\mathbf{a}| = \sqrt{a_1^2 + a_2^2}$.

For $\mathbf{a} = \langle a_1, a_2, a_3 \rangle$ the *length of* **a** is
$|\mathbf{a}| = \sqrt{a_1^2 + a_2^2 + a_3^2}$.

A *unit vector* has length 1. The *zero vector* is $\mathbf{0} = \langle 0, 0 \rangle$ or $\mathbf{0} = \langle 0, 0, 0 \rangle$.

In two dimensions $\mathbf{i} = \langle 1, 0 \rangle$ and $\mathbf{j} = \langle 0, 1 \rangle$ are the *standard basis vectors*.

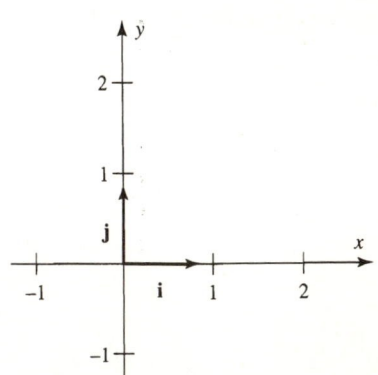

For any $\mathbf{a} = \langle a_1, a_2 \rangle$, we can write $\mathbf{a} = a_1\mathbf{i} + a_2\mathbf{j}$.
(We will have more to say about vector addition in part B.)

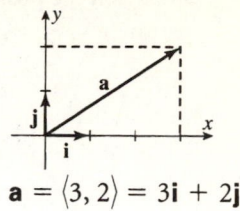

$$\mathbf{a} = \langle 3, 2 \rangle = 3\mathbf{i} + 2\mathbf{j}$$

In three-dimensional space the *standard basic vectors* are
$\mathbf{i} = \langle 1, 0, 0 \rangle$, $\mathbf{j} = \langle 0, 1, 0 \rangle$, and $\mathbf{k} = \langle 0, 0, 1 \rangle$.
Any $\mathbf{a} = \langle a_1, a_2, a_3 \rangle$ may be
written as $\mathbf{a} = a_1\mathbf{i} + a_2\mathbf{j} + a_3\mathbf{k}$.

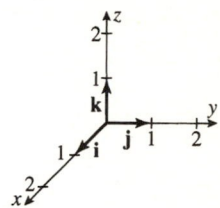

It will be very important to think of vectors in different, but equivalent ways. For
example, we will, as the need arises, think of the vector $\mathbf{a} = 2\mathbf{i} + 3\mathbf{j} + 1\mathbf{k}$ as

   i) an abstract object with a specified direction and a
      length ($\sqrt{14}$)

   ii) the point $(2, 3, 1)$ in space

   iii) a "pointy stick" with tail at the origin and head at $(2, 3, 1)$

   iv) a "pointy stick" with the same direction and magnitude as the one
      with tail at the origin, but its tail is not at the origin.

The ability to shift back and forth between such interpretations is useful in
understanding vector calculus. For example, in four dimensions with the
four unit basis vectors $\mathbf{u}_1, \mathbf{u}_2, \mathbf{u}_3, \mathbf{u}_4$, a vector $\mathbf{a} = \langle a_1, a_2, a_3, a_4 \rangle$ may be
written $\mathbf{a} = a_1\mathbf{u}_1 + a_2\mathbf{u}_2 + a_3\mathbf{u}_3 + a_4\mathbf{u}_4$ and $|\mathbf{a}| = \sqrt{a_1^2 + a_2^2 + a_3^2 + a_4^2}$.
The set of all n-dimensional vectors is denoted V_n.

1) The vector represented by the line segment
from $(5, 6)$ to $(3, 7)$ is _____.

$$\langle 3 - 5, 7 - 6 \rangle = \langle -2, 1 \rangle.$$

2) The vector represented by the directed line
segment from $(10, 4, 3)$ to $(5, 4, 6)$ is _____.

$$\langle 5 - 10, 4 - 4, 6 - 3 \rangle = \langle -5, 0, 3 \rangle.$$

3) Sketch the position vectors for each:

a) $\mathbf{a} = \langle 2, -6 \rangle$.

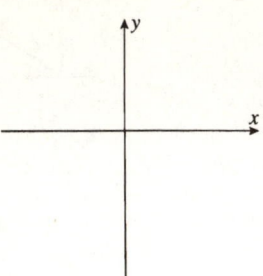

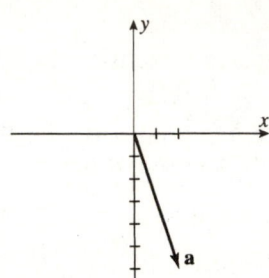

b) $\mathbf{a} = \langle 3, 6, 4 \rangle$.

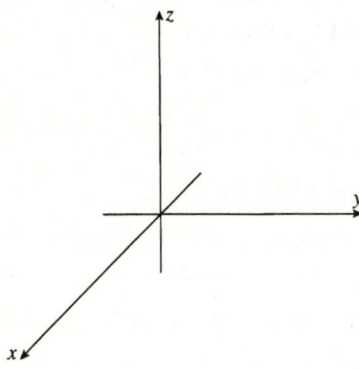

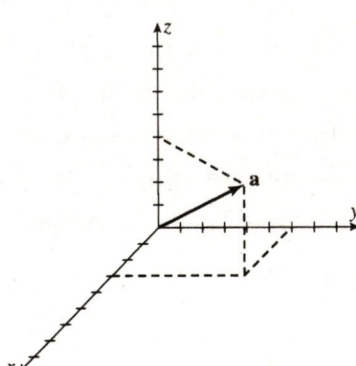

4) The length of $\mathbf{a} = \langle 4, 2, -2 \rangle$ is _____.

$$|\mathbf{a}| = \sqrt{4^2 + 2^2 + (-2)^2} = \sqrt{24}.$$

5) Is $\mathbf{a} = \left\langle \dfrac{-1}{4}, \dfrac{\sqrt{3}}{2}, \dfrac{\sqrt{3}}{4} \right\rangle$ a unit vector?

Yes, since

$$|\mathbf{a}| = \sqrt{\left(\dfrac{-1}{4}\right)^2 + \left(\dfrac{\sqrt{3}}{2}\right)^2 + \left(\dfrac{\sqrt{3}}{4}\right)^2} = 1.$$

6) Write each in terms of standard basis vectors:

    a) $\mathbf{a} = \langle 4, -3 \rangle$.

$$\mathbf{a} = 4\mathbf{i} - 3\mathbf{j}$$

    b) $\mathbf{a} = \langle 0, 4, 5 \rangle$.

$$\mathbf{a} = 4\mathbf{j} + 5\mathbf{k}$$

    c) $\mathbf{a}$ is represented by $\overrightarrow{AB}$ where $A: (2, 5)$, $B: (6, 1)$.

$$\mathbf{a} = \langle 6 - 2, 1 - 5 \rangle = \langle 4, -4 \rangle = 4\mathbf{i} - 4\mathbf{j}$$

Page 837 (ET Page 801)

B. Vector addition and scalar multiplication are done componentwise: For $\mathbf{a} = \langle a_1, a_2 \rangle$ and $\mathbf{b} = \langle b_1, b_2 \rangle$ and real number c (called a *scalar*),

$$\mathbf{a} + \mathbf{b} = \langle a_1 + b_1, a_2 + b_2 \rangle$$
$$c\,\mathbf{a} = \langle ca_1, ca_2 \rangle.$$

Graphically, $\mathbf{a} + \mathbf{b}$ is the vector obtained by positioning the initial point of $\mathbf{b}$ at the terminal point of $\mathbf{a}$ and drawing the directed line segment from the initial point of $\mathbf{a}$ to the terminal point of $\mathbf{b}$.

Another way to visualize $\mathbf{a} + \mathbf{b}$ is to place the tails of $\mathbf{a}$ and $\mathbf{b}$ together and draw the resulting parallelogram. $\mathbf{a} + \mathbf{b}$ is the diagonal of the parallelogram.

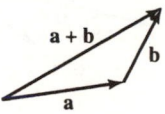
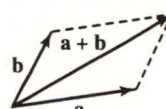

For $c > 0$, $c\,\mathbf{a}$ is the vector in the same direction as $\mathbf{a}$ with length c times the length of $\mathbf{a}$. For $c < 0$, $c\,\mathbf{a}$ is the vector in the opposite direction of $\mathbf{a}$ with length $-c$ times the length of $\mathbf{a}$.

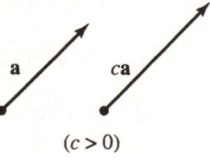
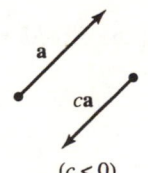

These definitions allow us to write $\mathbf{a} = \langle a_1, a_2 \rangle$ as $\mathbf{a} = a_1\mathbf{i} + a_2\mathbf{j}$ because $\langle a_1, a_2 \rangle = \langle a_1, 0 \rangle + \langle 0, a_2 \rangle = a_1\langle 1, 0 \rangle + a_2\langle 0, 1 \rangle = a_1\mathbf{i} + a_2\mathbf{j}$.

$\mathbf{a} - \mathbf{b}$ is defined to be $\mathbf{a} + (-\mathbf{b}) = \mathbf{a} + (-1)\mathbf{b}$.

For any nonzero vector $\mathbf{a}$, $\dfrac{\mathbf{a}}{|\mathbf{a}|}$ is a unit vector. Nonzero vector $\mathbf{a}$ is parallel to $\mathbf{b}$ means $\mathbf{b} = c\,\mathbf{a}$ for some c.

Properties of vectors:

    $\mathbf{a} + \mathbf{b} = \mathbf{b} + \mathbf{a}$ $c(\mathbf{a} + \mathbf{b}) = c\,\mathbf{a} + c\,\mathbf{b}$

    $\mathbf{a} + (\mathbf{b} + \mathbf{c}) = (\mathbf{a} + \mathbf{b}) + \mathbf{c}$ $(c + d)\mathbf{a} = c\,\mathbf{a} + d\,\mathbf{a}$

    $\mathbf{a} + \mathbf{0} = \mathbf{a}$ ($\mathbf{0}$ is the zero vector.) $(cd)\mathbf{a} = c(d\,\mathbf{a})$

Vector operations for three or more dimensions are defined similarly.

7) For $\mathbf{a} = \langle 4, 3 \rangle$ and $\mathbf{b} = \langle -5, 1 \rangle$:

a) $\mathbf{a} + \mathbf{b} = $ _____.

$\langle 4, 3 \rangle + \langle -5, 1 \rangle = \langle -1, 4 \rangle.$

b) $6\mathbf{a} = $ _____.

$\langle 6(4), 6(3) \rangle = \langle 24, 18 \rangle.$

c) Sketch the position vector of $\mathbf{b} - \mathbf{a}$.

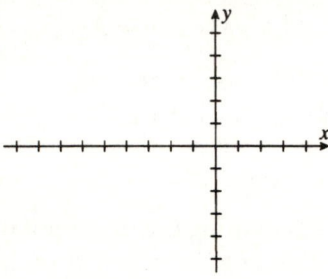

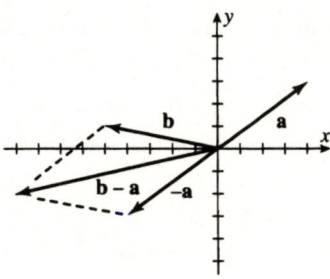

8) True or False:

$c(\mathbf{a} - \mathbf{b}) = c\mathbf{a} - c\mathbf{b}.$

True.

9) Find the unit vector in the direction of $\mathbf{a} = \langle 4, 3, -1 \rangle$.

$|\mathbf{a}| = \sqrt{4^2 + 3^2 + (-1)^2} = \sqrt{26}.$

The unit vector is $\dfrac{\mathbf{a}}{|\mathbf{a}|} = \left\langle \dfrac{4}{\sqrt{26}}, \dfrac{3}{\sqrt{26}}, \dfrac{-1}{\sqrt{26}} \right\rangle.$

10) For $\mathbf{a} = 7\mathbf{i} + 4\mathbf{j} - 3\mathbf{k}$ and $\mathbf{b} = 2\mathbf{i} + \mathbf{j} + 5\mathbf{k}$:

a) $\mathbf{a} + \mathbf{b} = $ _____.

$9\mathbf{i} + 5\mathbf{j} + 2\mathbf{k}$

b) $-2\mathbf{b} = $ _____.

$-4\mathbf{i} - 2\mathbf{j} - 10\mathbf{k}$

11) Is $\mathbf{a} = \langle 4, -8, 6 \rangle$ parallel to $\mathbf{b} = \langle -2, 3, -4 \rangle$?

No. Comparing first components $\mathbf{a} = -2\mathbf{b}$ if $\mathbf{a}$ and $\mathbf{b}$ are parallel. But comparing second components, $\mathbf{a} = -\frac{8}{3}\mathbf{b}$.

12) A plane is heading north at 250 mph but there is a 40 mph crosswind pushing the plane from the west. What is the true direction the plane is heading?

Let $\mathbf{u}$ be the vector representing due north at 250 mph. Let $\mathbf{v}$ be the wind from the west. The true heading is
$\mathbf{u} + \mathbf{v} = \langle 0, 250 \rangle + \langle 40, 0 \rangle = \langle 40, 250 \rangle$.

For the angle θ from north
$\tan \theta = \frac{40}{250} = 0.16$
$\theta = \tan^{-1} 0.16 = 0.159$
The true course is 0.159 radians ($\approx 9°$) east of due north.

Section 13.3 The Dot Product

The previous section showed you how to add vectors and stretch them (scalar multiplication). This section gives one way to multiply vectors with the result being a scalar. The dot product defined in this section provides a convienent way to determine the angle between two vectors.

Concepts to Master

A. Dot product; properties of $\mathbf{a} \cdot \mathbf{b}$

B. Angle between two vectors; Scalar projection; Work; Direction cosines

Summary and Focus Questions

Page 843 (ET Page 807)

A. For $\mathbf{a} = \langle a_1, a_2 \rangle$ and $\mathbf{b} = \langle b_1, b_2 \rangle$, the *dot product* is $\mathbf{a} \cdot \mathbf{b} = a_1 b_1 + a_2 b_2$.

For $\mathbf{a} = \langle a_1, a_2, a_3 \rangle$ and $\mathbf{b} = \langle b_1, b_2, b_3 \rangle$, the *dot product* $\mathbf{a} \cdot \mathbf{b}$ is
$\mathbf{a} \cdot \mathbf{b} = a_1 b_1 + a_2 b_2 + a_3 b_3$.

Properties of the dot product include
$$\mathbf{a} \cdot \mathbf{b} = \mathbf{b} \cdot \mathbf{a}$$
$$\mathbf{a} \cdot (\mathbf{b} + \mathbf{c}) = \mathbf{a} \cdot \mathbf{b} + \mathbf{a} \cdot \mathbf{c}$$
$$c(\mathbf{a} \cdot \mathbf{b}) = (c\mathbf{a}) \cdot \mathbf{b}$$
$$\mathbf{a} \cdot \mathbf{a} = |\mathbf{a}|^2$$

1) Find $\mathbf{a} \cdot \mathbf{b}$ for

a) $\mathbf{a} = \langle 6, 3 \rangle, \mathbf{b} = \langle 2, -1 \rangle$

$\mathbf{a} \cdot \mathbf{b} = 6(2) + 3(-1) = 9.$

b) $\mathbf{a} = 4\mathbf{i} - 3\mathbf{j} + \mathbf{k}, \mathbf{b} = 5\mathbf{j} + 10\mathbf{k}$

$\mathbf{a} \cdot \mathbf{b} = 4(0) + (-3)5 + 1(10) = -5.$

c) $\mathbf{a} = \langle 8, 1, 4 \rangle, \mathbf{b} = \langle 3, 0, -6 \rangle.$

$\mathbf{a} \cdot \mathbf{b} = 8(3) + 1(0)\,4(-6) = 0.$

2) Why is $\mathbf{a} \cdot \mathbf{b} \cdot \mathbf{c}$ not defined although $(\mathbf{a} \cdot \mathbf{b})\mathbf{c}$ is defined?

$\mathbf{a} \cdot \mathbf{b} \cdot \mathbf{c}$ makes no sense because the dot product $\mathbf{a} \cdot \mathbf{b}$ is scalar and the dot product of a scalar with a vector $\mathbf{c}$ is not defined. But $(\mathbf{a} \cdot \mathbf{b})\mathbf{c}$ [with no dot between the) and $\mathbf{c}$] is a scalar multiple of $\mathbf{c}$.

3) True or False: $\mathbf{a} \cdot \mathbf{a} = 1$ if and only if $\mathbf{a}$ is a unit vector.

True.

Page 844
(ET Page 808)

B. The *angle θ between vectors **a** and **b*** satisfies
$\mathbf{a} \cdot \mathbf{b} = |\mathbf{a}|\,|\mathbf{b}| \cos\theta$, or equivalently, $\cos\theta = \dfrac{\mathbf{a} \cdot \mathbf{b}}{|\mathbf{a}|\,|\mathbf{b}|}$.
a is *orthogonal* (*perpendicular*) to **b** iff $\mathbf{a} \cdot \mathbf{b} = 0$.

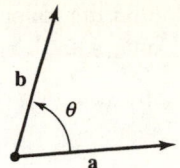

The *scalar projection* of **b** onto **a** is the number
$|\mathbf{b}| \cos\theta$. Intuitively, the scalar projection is the
length of the shadow of **b** cast upon **a** by a light
source directly over **a**.

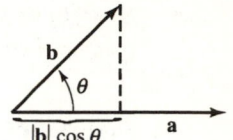

The *work* done by a constant force **F** moving an
object along a directed line segment represented
by a vector **a** is $\mathbf{F} \cdot \mathbf{a}$.

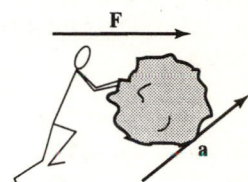

For nonzero $\mathbf{a} = (a_1, a_2, a_3)$, let
$\quad\alpha$ = angle between **a** and positive x-axis,
$\quad\beta$ = angle between **a** and positive y-axis,
$\quad\gamma$ = angle between **a** and positive z-axis.
$\cos\alpha = \dfrac{a_1}{|\mathbf{a}|}$, $\cos\beta = \dfrac{a_2}{|\mathbf{a}|}$, $\cos\gamma = \dfrac{a_3}{|\mathbf{a}|}$ are the
direction cosines for the direction angles α, β,
and γ.

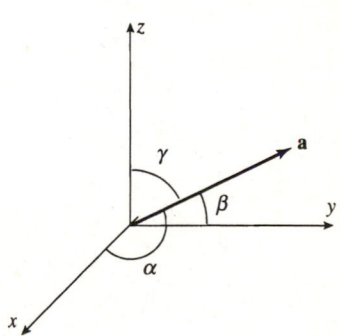

4) Find the angle θ between the vectors
$\mathbf{a} = \langle 1, -1, -1 \rangle$ and $\mathbf{b} = \langle -1, 5, 1 \rangle$.

$$\cos\theta = \frac{\mathbf{a} \cdot \mathbf{b}}{|\mathbf{a}|\,|\mathbf{b}|}$$

$$= \frac{1(-1) + (-1)5 + (-1)1}{\sqrt{1^2 + (-1)^2 + (-1)^2}\,\sqrt{(-1)^2 + 5^2 + 1^2}}$$

$$= \frac{-7}{9}.$$

Therefore,
$$\theta = \cos^{-1}\!\left(\frac{-7}{9}\right) \approx 2.46 \text{ radians} \approx 141°.$$

5) Find the scalar projection of $\mathbf{x} = 3\mathbf{i} + \mathbf{j} + \mathbf{k}$ onto $\mathbf{y} = 2\mathbf{i} + 3\mathbf{j} - 4\mathbf{k}$.

$|\mathbf{x}| = \sqrt{11}$, $|\mathbf{y}| = \sqrt{29}$. For θ, the angle between $\mathbf{x}$ and $\mathbf{y}$, $\cos \theta = \dfrac{\mathbf{x} \cdot \mathbf{y}}{|\mathbf{x}|\,|\mathbf{y}|} = \dfrac{5}{\sqrt{11}\,\sqrt{29}}$.

The projection is
$|\mathbf{x}| \cos \theta = \sqrt{11}\,\dfrac{5}{\sqrt{11}\,\sqrt{29}} = \dfrac{5}{\sqrt{29}}$.

6) Is $6\mathbf{i} - 3\mathbf{j} + 2\mathbf{k}$ orthogonal to $2\mathbf{i} + 10\mathbf{j} + 9\mathbf{k}$?

Yes. The dot product is
$6(2) + (-3)10 + 2(9) = 0$.

7) Find the amount of work done by a horizontal force of 4 newtons moving an object 3 meters up a hill that makes a $30°$ angle with the horizontal ground. (Disregard gravity.)

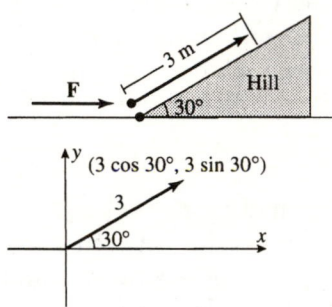

The vector representing the distance travelled is
$\mathbf{d} = \langle 3 \cos 30°, 3 \sin 30° \rangle = \left\langle \dfrac{3\sqrt{3}}{2}, \dfrac{3}{2} \right\rangle$.

The force $\mathbf{F} = \langle 4, 0 \rangle$ so the work done is
$\mathbf{F} \cdot \mathbf{d} = 4\left(\dfrac{3\sqrt{3}}{2}\right) + 0\left(\dfrac{3}{2}\right) = 6\sqrt{3}$ joules.

8) Find the direction cosines of
$\mathbf{a} = 3\mathbf{i} - \mathbf{j} + 4\mathbf{k}$.

$|\mathbf{a}| = \sqrt{3^2 + (-1)^2 + 4^2} = \sqrt{26}$.
$\cos \alpha = \dfrac{3}{\sqrt{26}}$, $\cos \beta = \dfrac{-1}{\sqrt{26}}$, $\cos \gamma = \dfrac{4}{\sqrt{26}}$.

9) Find any nonzero vector **x** orthogonal to
$\mathbf{y} = \langle 3, 4 \rangle$.

If $\mathbf{x} = \langle a, b \rangle$ is orthogonal to y, then
$\mathbf{x} \cdot \mathbf{y} = 3a + 4b = 0$. Any solution to this
equation other than $(0, 0)$ will do. Let $a = 4$
and $b = -3$. Then $\mathbf{x} = \langle 4, -3 \rangle$ is
orthogonal to **y**.

10) Find a unit vector orthogonal to
$\mathbf{a} = \langle 4, -3, 6 \rangle$.

First, find a vector $\mathbf{b} = \langle x, y, z \rangle$ where
$\mathbf{a} \cdot \mathbf{b} = 0$. $\mathbf{a} \cdot \mathbf{b} = 4x - 3y + 6z = 0$. One
of infinitely many solutions is $x = 3$,
$y = 4, z = 0$; thus $\mathbf{b} = \langle 3, 4, 0 \rangle$.
$|\mathbf{b}| = \sqrt{3^2 + 4^2 + 0^2} = 5$.
The unit vector is $\dfrac{\mathbf{b}}{|\mathbf{b}|} = \left\langle \dfrac{3}{5}, \dfrac{4}{5}, 0 \right\rangle$.

Section 13.4   The Cross Product

This section provides a second way to multiply two three-dimensional vectors – this time producing a third vector perpendicular to the given vectors. Cross product is defined only for three-dimensional vectors and has many uses; we shall see, for example, in the next section that the cross product of two vectors is useful for determining the equation of the plane formed by those vectors.

Concepts to Master

A. Cross product; Properties of **a** × **b**

B. Area of parallelogram; Scalar triple product; Volume of parallelepiped

Summary and Focus Questions

Page 850 (ET Page 814)

A. The *cross product* **a** × **b** of vectors $\mathbf{a} = \langle a_1, a_2, a_3 \rangle$ and $\mathbf{b} = \langle b_1, b_2, b_3 \rangle$ is $\mathbf{a} \times \mathbf{b} = \langle a_2 b_3 - a_3 b_2, a_3 b_1 - a_1 b_3, a_1 b_2 - a_2 b_1 \rangle$.

Calculating **a** × **b** may be done using determinants:
For a two by two array (a square of four numbers), the *determinant* is:

$$\begin{vmatrix} x & y \\ z & w \end{vmatrix} = xw - zy,$$

which may be remembered as the product of the numbers on the upper left to lower right "diagonal" (xw) minus the product from the lower left to the upper right "diagonal" (zy).

For example $\begin{vmatrix} 6 & 2 \\ 3 & 5 \end{vmatrix} = 6(5) - 3(2) = 30 - 6 = 24$.

The determinant of a three by three square array of 9 numbers may be used to remember how to compute **a** × **b**: use vectors **i, j, k** as symbols in the first row of the determinant, the components of **a** in the second row and the components of **b** in the third row.

For example, if $\mathbf{a} = \langle 2, 1, 5 \rangle$ and $\mathbf{b} = \langle 4, 6, 3 \rangle$, then

$$\mathbf{a} \times \mathbf{b} = \begin{vmatrix} \mathbf{i} & \mathbf{j} & \mathbf{k} \\ 2 & 1 & 5 \\ 4 & 6 & 3 \end{vmatrix}.$$

This is evaluated as follows:
the determinant obtained by deleting the row and column of **i** times **i**

$$\begin{vmatrix} 1 & 5 \\ 6 & 3 \end{vmatrix} \mathbf{i} = (1 \cdot 3 - 6 \cdot 5)\mathbf{i} = -27\mathbf{i}$$

minus
the determinant obtained by deleting the row and column of **j** times **j**

$$\begin{vmatrix} 2 & 5 \\ 4 & 3 \end{vmatrix}\mathbf{j} = (2 \cdot 3 - 4 \cdot 5)\mathbf{j} = -14\mathbf{j}$$

plus
the determinant obtained by deleting the row and column of **k** times **k**.

$$+ \begin{vmatrix} 2 & 1 \\ 4 & 6 \end{vmatrix}\mathbf{k} = (2 \cdot 6 - 4 \cdot 1)\mathbf{k} = 8\mathbf{k}$$

Thus $\mathbf{a} \times \mathbf{b} = -27\mathbf{i} - (-14\mathbf{j}) + 8\mathbf{k} = -27\mathbf{i} + 14\mathbf{j} + 8\mathbf{k}$.

The direction of $\mathbf{a} \times \mathbf{b}$ follows the "right hand rule": if the fingers of your right hand curve in the direction of the angle from **a** to **b**, then your thumb points in the direction of $\mathbf{a} \times \mathbf{b}$.

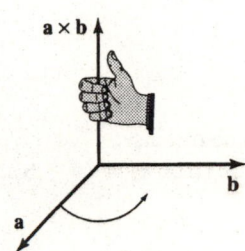

Properties of cross products include:

$\mathbf{a} \times \mathbf{b}$ is orthogonal to both **a** and **b**.
$|\mathbf{a} \times \mathbf{b}| = |\mathbf{a}|\,|\mathbf{b}| \sin \theta, \theta = $ the angle between **a** and **b**.
$\mathbf{a} \times \mathbf{b} = -\mathbf{b} \times \mathbf{a}$
$c(\mathbf{a} \times \mathbf{b}) = c\mathbf{a} \times \mathbf{b} = \mathbf{a} \times c\mathbf{b}$ (c, a scalar)
$\mathbf{a} \times (\mathbf{b} + \mathbf{c}) = \mathbf{a} \times \mathbf{b} + \mathbf{a} \times \mathbf{c}$ and $(\mathbf{a} + \mathbf{b}) \times \mathbf{c} = \mathbf{a} \times \mathbf{c} + \mathbf{b} \times \mathbf{c}$.
$\mathbf{a} \times (\mathbf{b} \times \mathbf{c}) = (\mathbf{a} \cdot \mathbf{c})\mathbf{b} - (\mathbf{a} \cdot \mathbf{b})\mathbf{c}$
$\mathbf{i} \times \mathbf{j} = \mathbf{k},\ \mathbf{j} \times \mathbf{k} = \mathbf{i},\ $ and $\mathbf{k} \times \mathbf{i} = \mathbf{j}$.

1) Find $\mathbf{a} \times \mathbf{b}$, where $\mathbf{a} = \langle 4, -1, 3 \rangle$ and $\mathbf{b} = \langle 1, 6, 2 \rangle$.

$$\mathbf{a} \times \mathbf{b} = \begin{vmatrix} \mathbf{i} & \mathbf{j} & \mathbf{k} \\ 4 & -1 & 3 \\ 1 & 6 & 2 \end{vmatrix}$$

$$= \begin{vmatrix} -1 & 3 \\ 6 & 2 \end{vmatrix}\mathbf{i} - \begin{vmatrix} 4 & 3 \\ 1 & 2 \end{vmatrix}\mathbf{j} + \begin{vmatrix} 4 & -1 \\ 1 & 6 \end{vmatrix}\mathbf{k}$$

$$= -20\mathbf{i} - 5\mathbf{j} + 25\mathbf{k}.$$

2) For any vectors **a**, **b** what is $\mathbf{a} \times \mathbf{b} + \mathbf{b} \times \mathbf{a}$?

Since $\mathbf{a} \times \mathbf{b} = -\mathbf{b} \times \mathbf{a}$, $\mathbf{a} \times \mathbf{b} + \mathbf{b} \times \mathbf{a} = 0$, the zero vector.

3) Evaluate $2\mathbf{i} \times (4\mathbf{j} + 3\mathbf{k})$.

This may be done using determinants but we use cross product properties:

$$2\mathbf{i} \times (4\mathbf{j} + 3\mathbf{k}) = 2\mathbf{i} \times 4\mathbf{j} + 2\mathbf{i} \times 3\mathbf{k}$$
$$= 8(\mathbf{i} \times \mathbf{j}) + 6(\mathbf{i} \times \mathbf{k})$$
$$= 8(\mathbf{i} \times \mathbf{j}) - 6(\mathbf{k} \times \mathbf{i})$$
$$= 8\mathbf{k} - 6\mathbf{j}.$$

4) Find a vector orthogonal to both $2\mathbf{i} - 3\mathbf{j} + \mathbf{k}$ and $\mathbf{i} + \mathbf{j} + 2\mathbf{k}$.

The cross product will do:

$$\begin{vmatrix} \mathbf{i} & \mathbf{j} & \mathbf{k} \\ 2 & -3 & 1 \\ 1 & 1 & 2 \end{vmatrix} = \begin{vmatrix} -3 & 1 \\ 1 & 2 \end{vmatrix}\mathbf{i} - \begin{vmatrix} 2 & 1 \\ 1 & 2 \end{vmatrix}\mathbf{j} + \begin{vmatrix} 2 & -3 \\ 1 & 1 \end{vmatrix}\mathbf{k}$$

$$= -7\mathbf{i} - 3\mathbf{j} + 5\mathbf{k}.$$

5) True or False:
$\mathbf{a} \times (\mathbf{b} \times \mathbf{c}) = (\mathbf{a} \times \mathbf{b}) \times \mathbf{c}$.

False. For example, let $\mathbf{a} = \langle 1, 2, 3 \rangle$, $\mathbf{b} = \langle 4, 5, 6 \rangle$ and $\mathbf{c} = \langle 7, 8, 9 \rangle$. Then $\mathbf{a} \times (\mathbf{b} \times \mathbf{c}) = \langle -24, -6, 12 \rangle$ while $(\mathbf{a} \times \mathbf{b}) \times \mathbf{c} = \langle 78, 6, -66 \rangle$.

B. $|\mathbf{a} \times \mathbf{b}|$ is the area of the parallelogram formed by $\mathbf{a}$ and $\mathbf{b}$.

Page 853 (ET Page 817)

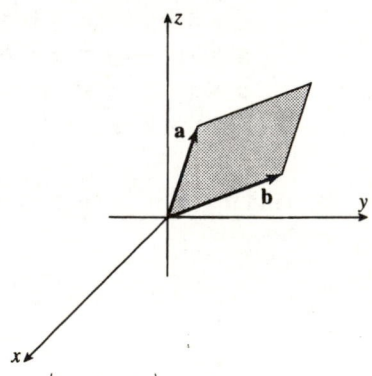

The triple scalar product of vectors $\mathbf{a} = \langle a_1, a_2, a_3 \rangle$, $\mathbf{b} = \langle b_1, b_2, b_3 \rangle$, and $\mathbf{c} = \langle c_1, c_2, c_3 \rangle$ is the number $\mathbf{a} \cdot (\mathbf{b} \times \mathbf{c})$. It may be evaluated directly or by using

$$\mathbf{a} \cdot (\mathbf{b} \times \mathbf{c}) = \begin{vmatrix} a_1 & a_2 & a_3 \\ b_1 & b_2 & b_3 \\ c_1 & c_2 & c_3 \end{vmatrix}.$$

$|\mathbf{a} \cdot (\mathbf{b} \times \mathbf{c})|$ is the volume of the parallelepiped formed by $\mathbf{a}$, $\mathbf{b}$, and $\mathbf{c}$.

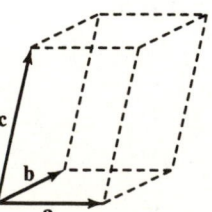

6) Find the area of the parallelogram with vertices P: $(4, 1, 3)$, Q: $(7, 5, 3)$, R: $(6, 4, 2)$, S: $(9, 8, 2)$.

Let $\mathbf{a} = \overrightarrow{PQ} = \langle 7 - 4, 5 - 1, 3 - 3 \rangle$
$= \langle 3, 4, 0 \rangle$.
Let $\mathbf{b} = \overrightarrow{PR} = \langle 6 - 4, 4 - 1, 2 - 3 \rangle$
$= \langle 2, 3, -1 \rangle$.
The area we seek is $|\mathbf{a} \times \mathbf{b}|$.

$$\mathbf{a} \times \mathbf{b} = \begin{vmatrix} \mathbf{i} & \mathbf{j} & \mathbf{k} \\ 3 & 4 & 0 \\ 2 & 3 & -1 \end{vmatrix} = -4\mathbf{i} + 3\mathbf{j} + \mathbf{k}.$$

Thus $|\mathbf{a} \times \mathbf{b}| = \sqrt{(-4)^2 + 3^2 + (-1)^2} = \sqrt{26}$.

7) Find $\mathbf{a} \cdot (\mathbf{b} \times \mathbf{c})$ where $\mathbf{a} = \langle 1, 3, 1 \rangle$, $\mathbf{b} = \langle 0, 4, 2 \rangle$, $\mathbf{c} = \langle -2, 2, 3 \rangle$:

a) directly

$$\mathbf{b} \times \mathbf{c} = \begin{vmatrix} \mathbf{i} & \mathbf{j} & \mathbf{k} \\ 0 & 4 & 2 \\ -2 & 2 & 3 \end{vmatrix}$$
$$= \begin{vmatrix} 4 & 2 \\ 2 & 3 \end{vmatrix}\mathbf{i} - \begin{vmatrix} 0 & 2 \\ -2 & 3 \end{vmatrix}\mathbf{j} + \begin{vmatrix} 0 & 4 \\ -2 & 2 \end{vmatrix}\mathbf{k}$$
$$= 8\mathbf{i} - 4\mathbf{j} + 8\mathbf{k}.$$
Thus $\mathbf{a} \cdot (\mathbf{b} \times \mathbf{c}) = 1(8) + 3(-4) + 1(8)$
$= 4.$

b) using determinants

$$\mathbf{a} \cdot (\mathbf{b} \times \mathbf{c}) = \begin{vmatrix} 1 & 3 & 1 \\ 0 & 4 & 2 \\ -2 & 2 & 3 \end{vmatrix}$$
$$= 1\begin{vmatrix} 4 & 2 \\ 2 & 3 \end{vmatrix} - 3\begin{vmatrix} 0 & 2 \\ -2 & 3 \end{vmatrix} + 1\begin{vmatrix} 0 & 4 \\ -2 & 2 \end{vmatrix}$$
$$= 1(8) - 3(4) + 1(8) = 4.$$

8) Find the volume of the parallelepiped formed by $\mathbf{a} = -2\mathbf{i} + 3\mathbf{j}$, $\mathbf{b} = 4\mathbf{i} + \mathbf{j} - \mathbf{k}$, $\mathbf{c} = 6\mathbf{j} + \mathbf{k}$.

$$\mathbf{a} \cdot (\mathbf{b} \times \mathbf{c}) = \begin{vmatrix} -2 & 3 & 0 \\ 4 & 1 & -1 \\ 0 & 6 & 1 \end{vmatrix}$$
$$= -2\begin{vmatrix} 1 & -1 \\ 6 & 1 \end{vmatrix} - 3\begin{vmatrix} 4 & -1 \\ 0 & 1 \end{vmatrix} + 0\begin{vmatrix} 4 & 1 \\ 0 & 6 \end{vmatrix}$$
$$= -2(7) - 3(4) + 0 = -26.$$
$|\mathbf{a} \cdot (\mathbf{b} \times \mathbf{c})| = 26.$

Section 13.5   Equations of Lines and Planes

This section gives three different, but equivalent, "equations" for a line in three-dimensional space – a vector form, a parametric form, and a form similar to that of a line in two dimensions. Planes in three-dimensional space, which may be thought of as all vectors perpendicular to a given vector, will have an equation form very similar to that of lines in two dimensions. We will use equations of lines and planes in later chapters when we discuss tangent planes to surfaces – a higher dimension version of tangent lines to graphs.

Concepts to Master

A. Equations of lines in space (vector, parametric, and symmetric); Parallel and skew lines

B. Normal vector; Equations of planes (vector and scalar); Parallel planes; Angle between intersecting planes; Distance from a point to a plane and distance between two planes

Summary and Focus Questions

**Page 858
(ET Page 822)**

A. Let L be a line passing through P_0: (x_0, y_0, z_0) and parallel to vector $\mathbf{v} = \langle a, b, c \rangle$. Let $\mathbf{r}_0 = \langle x_0, y_0, z_0 \rangle$. The three forms of the equations for L are

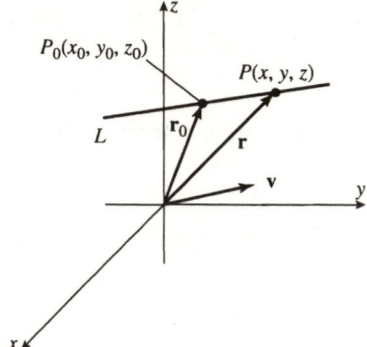

Vector Form	Parametric Form	Symmetric Form
$\mathbf{r} = \mathbf{r}_0 + t\mathbf{v}$	$x = x_0 + at$ $y = y_0 + bt$ $z = z_0 + ct$	$\dfrac{x - x_0}{a} = \dfrac{y - y_0}{b} = \dfrac{z - z_0}{c}$ a, b, c nonzero

In case, for example, $b = 0$, the symmetric form is modified to
$y = y_0$ and $\dfrac{x - x_0}{a} = \dfrac{z - z_0}{c}$.

For lines L_1: $\mathbf{r} = \mathbf{r}_0 + t_1\mathbf{v}_1$ and L_2: $\mathbf{r} = \mathbf{s}_0 + t_2\mathbf{v}_2$
   1) L_1 and L_2 are parallel iff $\mathbf{v}_1$ is a scalar multiple of $\mathbf{v}_2$.
   2) L_1 and L_2 intersect iff $\mathbf{r}_0 + t_1\mathbf{v}_1 = \mathbf{s}_0 + t_2\mathbf{v}_2$ for some t_1, t_2.

Lines that are not parallel and do not intersect are called *skew*.

1) Find the three forms for the equation of the line L through $(4, 1, 6)$ parallel to $\langle 7, 3, 0 \rangle$.

Vector form: $\mathbf{r} = \langle 4, 1, 6 \rangle + t\langle 7, 3, 0 \rangle$, or $\mathbf{r} = (4 + 7t)\mathbf{i} + (1 + 3t)\mathbf{j} + 6\mathbf{k}$.

Parametric form: $x = 4 + 7t$
$$y = 1 + 3t$$
$$z = 6 + 0t \ (\text{or } z = 6).$$

Symmetric form: $z = 6$ and $\dfrac{x - 4}{7} = \dfrac{y - 1}{3}$.

2) Determine whether the lines L_1 and L_2 are parallel, intersect, or are skew:

L: $x = 1 + 5t$ L_2: $x = 10 + 3s$
 $y = 3 + 4t$ $y = 17 - s$
 $z = -2 + t$ $z = -3 + 2s$

The direction $\langle 5, 4, 1 \rangle$ is not a multiple of the direction $\langle 3, -1, 2 \rangle$ so L_1 and L_2 are not parallel.

If L_1 and L_2 intersect, then
$$1 + 5t = 10 + 3s$$
$$3 + 4t = 17 - s$$
$$-2 + t = -3 + 2s$$
must have a solution. Multiplying the second equation by 2 and adding to the third gives
$$4 + 9t = 31,$$
$$9t = 27, t = 3.$$

Using $t = 3$ in the third,
$-2 + 3 = -3 + 2s$. Thus $2s = 4, s = 2$.

The solution to the second and third equations is $t = 3, s = 2$. Since this also satisfies the first $[1 + 5(3) = 10 + 3(2)]$ the lines intersect. At $t = 3$ and $s = 2$, $x = 16$, $y = 15, z = 1$ so the point of intersection is $(16, 15, 1)$.

3) Find the equation of the line parallel to the line L through the point P. Use the same form as the equation for L.

a) L: $\dfrac{x - 1}{8} = \dfrac{y + 2}{5} = \dfrac{z + 3}{-2}$

 P: $(4, 5, 1)$

$$\frac{x - 4}{8} = \frac{y - 5}{5} = \frac{z - 1}{-2}.$$

b) L: $\mathbf{r} = (3 + 4t)\mathbf{i} - t\mathbf{j} + (8 + t)\mathbf{k}$
P: $(-1, 3, 2)$

$\mathbf{r} = (-1 + 4t)\mathbf{i} + (3 - t)\mathbf{j} + (2 + t)\mathbf{k}.$

c) L: $x = 9 - 9t$
$y = 8 + 3t$
$z = 2 + 4t$
P: $(7, -3, 5)$

$x = 7 - 9t.$
$y = -3 + 3t.$
$z = 5 + 4t.$

**Page
861
(ET Page
825)**

B. A vector orthogonal to a plane is called a *normal vector*.

Suppose a plane contains a point P: (x_0, y_0, z_0) and has a normal vector $\mathbf{n} = \langle a, b, c \rangle$.

Let $\mathbf{r}_0 = \langle x_0, y_0, z_0 \rangle$. The two forms of the equation of the plane are

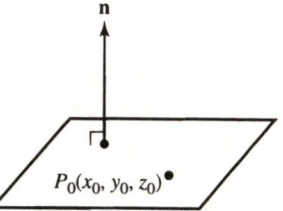

Vector Form	**Scalar Form**
$\mathbf{n}\mathbf{r} = \mathbf{n}\mathbf{r}_0$	$a(x - x_0) + b(y - y_0) + c(z - z_0) = 0$

If **a** and **b** are nonparallel vectors in a plane then $\mathbf{a} \times \mathbf{b}$ may be used as a normal vector for the plane.

Every linear equation of the form $Ax + By + Cz = D$ is the equation of a plane in three-dimensional space.

Two planes are parallel if their normal vectors are parallel.

The angle θ between two intersecting planes with normal vectors $\mathbf{n}_1$ and $\mathbf{n}_2$ satisfies

$$\cos \theta = \frac{\mathbf{n}_1 \cdot \mathbf{n}_2}{|\mathbf{n}_1| \, |\mathbf{n}_2|}.$$

The distance from a point P: (x_0, y_0, z_0) to a plane $ax + by + cz = d$ is

$$\frac{|ax_0 + by_0 + cz_0 - d|}{\sqrt{a^2 + b^2 + c^2}}.$$

This formula may also be used to find the distance between two parallel planes—just let P be a point on one of the planes.

4) Find the two forms of the equation of the plane through $(8, 2, 3)$ with normal vector $\langle 6, 2, 5 \rangle$.

Vector form:
$$\langle 6, 2, 5 \rangle \cdot \mathbf{r} = \langle 6, 2, 5 \rangle \cdot \langle 8, 2, 3 \rangle$$
$$\langle 6, 2, 5 \rangle \cdot \mathbf{r} = 67.$$

Scalar form:
$$6(x - 8) + 2(y - 2) + 5(z - 3) = 0$$
$$6x + 2y + 5z = 67.$$

5) Find the scalar equation of the plane through $P: (1, 3, 1)$, $Q: (3, 0, 4)$, $R: (4, -1, 2)$.

Let $\mathbf{a} = \overrightarrow{PQ} = \langle 2, -3, 3 \rangle$ and $\mathbf{b} = \overrightarrow{PR} = \langle 3, -4, 1 \rangle$.

A normal to the plane is
$$\mathbf{a} \times \mathbf{b} = \begin{vmatrix} \mathbf{i} & \mathbf{j} & \mathbf{k} \\ 2 & -3 & 3 \\ 3 & -4 & 1 \end{vmatrix} = 9\mathbf{i} + 7\mathbf{j} + \mathbf{k}.$$

Using P as a point on the plane the equation is
$$9(x - 1) + 7(y - 3) + (z - 1) = 0 \text{ or}$$
$$9x + 7y + z = 31.$$

6) Find the cosine of the angle of intersection of the planes:
$$3x + y + 3z = 5$$
$$6x - 3y - 2z = 14.$$

Let $\mathbf{n}_1 = \langle 3, 1, 3 \rangle$, $\mathbf{n}_2 = \langle 6, -3, -2 \rangle$.
$$\cos \theta = \frac{\mathbf{n}_1 \cdot \mathbf{n}_2}{|\mathbf{n}_1| \, |\mathbf{n}_2|} = \frac{9}{\sqrt{19}\sqrt{49}} = \frac{9}{7\sqrt{19}}.$$

7) Find the line of intersection of the planes:
$$2x + y - z = -4$$
$$3x + y + 2z = 5.$$

Let $\mathbf{n}_1 = \langle 2, 1, -1 \rangle$, $\mathbf{n}_2 = \langle 3, 1, 2 \rangle$.
The line of intersection has the same direction as
$$\mathbf{n}_1 \times \mathbf{n}_2 = \begin{vmatrix} \mathbf{i} & \mathbf{j} & \mathbf{k} \\ 2 & 1 & -1 \\ 3 & 1 & 2 \end{vmatrix} = \langle 3, -7, -1 \rangle.$$

To find a point of intersection set $x = 0$ ($y = 0$ or $z = 0$ also could be used):
$$y - z = -4$$
$$y + 2z = 5$$
Subtracting: $3z = 9$, $z = 3$.
Thus $y - 3 = -4$, $y = -1$.

The line of intersection is
$$\frac{x}{3} = \frac{y + 1}{-7} = \frac{z - 3}{-1}.$$

8) Find the distance from $(1, 2, -2)$ to
$x + 3y + 4z = 12$.

$$\frac{\left|1(1) + 3(2) + 4(-2) - 12\right|}{\sqrt{1^2 + 3^2 + 4^2}} = \frac{|-13|}{\sqrt{26}} = \frac{\sqrt{26}}{2}.$$

9) Find the distance between the parallel planes:
$2x + 3y + 7z = 10$ and
$2x + 3y + 7z = 134$.

$(5, 0, 0)$ is a point on the first plane. The distance between the planes is
$$\frac{\left|2(5) + 3(0) + 7(0) - 134\right|}{\sqrt{2^2 + 3^2 + 7^2}} = \frac{|-124|}{\sqrt{62}}$$
$$= \frac{124}{\sqrt{62}} = 2\sqrt{62}.$$

10) Where does the line $\dfrac{x - 1}{1} = \dfrac{y + 2}{3} = \dfrac{z - 1}{2}$
intersect the plane $2x - 3y + z = -6$?

First rewrite the line in parametric form:
$$x = 1 + t$$
$$y = -2 + 3t$$
$$z = 1 + 2t.$$

Now substitution into the equation of the plane:
$$2(1 + t) - 3(-2 + 3t) + (1 + 2t) = -6$$
$$9 - 5t = -6$$
$$-5t = -15$$
$$t = 3.$$

Thus $x = 1 + 3 = 4$, $y = -2 + 3(3) = 7$, and $z = 1 + 2(3) = 7$. The point of intersection is $(4, 7, 7)$.

Section 13.6 Cylinders and Quadric Surfaces

This section describes the equations and graphs of several types of surfaces in three dimensions; they are reminiscent of the conic sections (ellipses, parabolas, and hyperbolas) in two dimensions. We first start with cylinders, which consist of collections of parallel lines.

Concepts to Master

A. Cylinders in three-dimensional space
B. Non-degenerate quadric surfaces

Summary and Focus Questions

Page 868 (ET Page 832)

A. A *cylinder* is a surface composed of all lines parallel to a given line that pass through a given curve. A cylinder need not be a "round tube" as the common use of the term cylinder suggests. The plane $z = 2$, for example, fits the definition of a cylinder – all lines parallel to the y-axis through the line $L: x = t$, $y = 0, z = 2$.

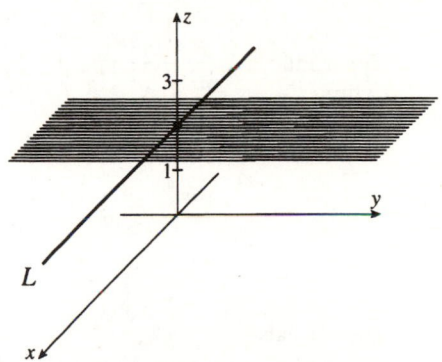

Some of the cylinders that are easiest to visualize are defined by equations where one or more of the variables (x, y, or z) is missing. Help in identifying a surface can come from examining the surface's *traces* – the intersections with a plane parallel to one of the xy-, xz-, or yz- planes. Here are some cylinders with conic sections as traces.

| | Typical | Traces | | | Typical |
Surface	Equation	$xy(z = 0)$	$xz(y = 0)$	$yz(x = 0)$	Graph
Parabolic Cylinder	$x^2 = 4ay$	parabola	z-axis	z-axis	
Elliptic Cylinder	$\dfrac{x^2}{a^2} + \dfrac{y^2}{b^2} = 1$	ellipse	two parallel lines	two parallel lines	
Hyperbolic Cylinder	$\dfrac{x^2}{a^2} - \dfrac{y^2}{b^2} = 1$	hyperbola	two parallel lines	none	

1) True or False:
If one of x, y, z is missing from an equation, the surface is a cylinder.

True

2) Identify each:

a) $x^2 - 5z^2 = 0$

Hyperbolic cylinder along the y-axis.

b)

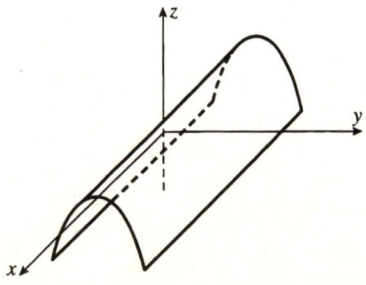

Parabolic cylinder.

3) Sketch a graph of each:

a) $x + 2y = 6$

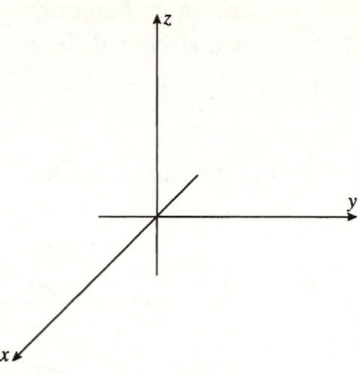

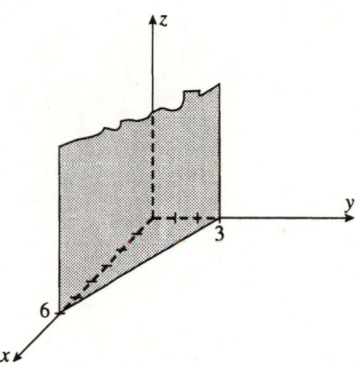

b) $\frac{x^2}{9} + \frac{z^2}{25} = 1$

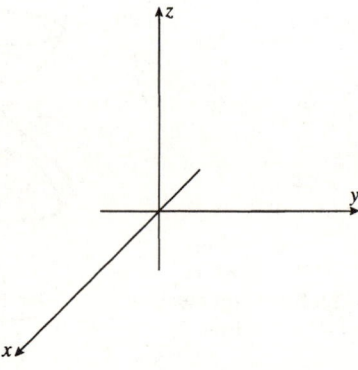

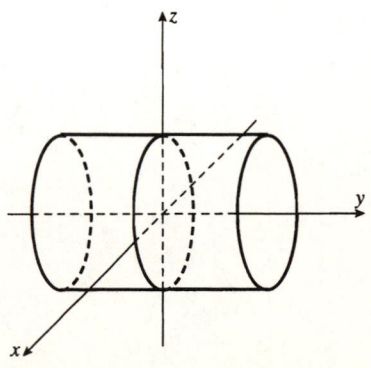

Page
869
(ET Page
833)

B. The analogues of the three conics in two dimensions are *quadric surfaces* in three dimensions. Again, it will help to first determine traces.

The table below summarizes the equations and their traces. Interchanging x, y, and z will produce surfaces of the same type but oriented along a different axis. For example $z^2 = \frac{x^2}{a^2} + \frac{y^2}{b^2}$, $y^2 = \frac{x^2}{a^2} + \frac{z^2}{c^2}$, and $x^2 = \frac{y^2}{b^2} + \frac{z^2}{c^2}$ are all elliptic cones.

Some equations may need to be rewritten by completing the square before trying to identify the surface.

Surface	Typical Equation	Traces $xy(z=0)$	$xz(y=0)$	$yz(x=0)$	Typical Graph
Ellipsoid	$\frac{x^2}{a^2} + \frac{y^2}{b^2} + \frac{z^2}{c^2} = 1$	ellipse	ellipse	ellipse	
Hyperboloid of One Sheet	$\frac{x^2}{a^2} + \frac{y^2}{b^2} - \frac{z^2}{c^2} = 1$	ellipse	hyperbola	hyperbola	
Hyperboloid of Two Sheets	$-\frac{x^2}{a^2} - \frac{y^2}{b^2} + \frac{z^2}{c^2} = 1$	none	hyperbola	hyperbola	
Elliptic Cone	$\frac{z^2}{c^2} = \frac{x^2}{a^2} + \frac{y^2}{b^2}$	$(0, 0, 0)$	two intersecting lines	two intersecting lines	

Surface	Typical Equation	Traces			Typical Graph
		$xy(z=0)$	$xz(y=0)$	$yz(x=0)$	
Elliptic Paraboloid	$\dfrac{z}{c}=\dfrac{x^2}{a^2}+\dfrac{y^2}{b^2}$	$(0,0,0)$	parabola upward	parabola upward	
Hyperbolic Paraboloid	$\dfrac{z}{c}=\dfrac{y^2}{b^2}-\dfrac{x^2}{a^2}$	two intersecting lines	parabola downward	parabola upward	

4) True or False:
A hyperbolic paraboloid must have an x^2 term, a y^2 term, and a z term.

False. Two variables must be squared and the third one is not, but they need not be $x^2, y^2,$ and z. For example, $y = \dfrac{x^2}{4} - \dfrac{z^2}{9}$ is a hyperbolic paraboloid.

5) Identify each:

a) $\dfrac{x^2}{25} - \dfrac{y^2}{4} - \dfrac{z^2}{16} = 1$

Hyperboloid of two sheets.

b) $y - 4x^2 = 16z^2$

Elliptic paraboloid.

c) $y^2 = x^2 + z^2$

Elliptic cone.

d) $x^2 - 8x - y^2 - 2y + z^2 + 15 = 0$

Complete the square:
$x^2 - 8x + 16 - (y^2 + 2y + 1) + z^2 = 0$
$(x - 4)^2 - (y + 1)^2 + z^2 = 0$
$(y + 1)^2 = (x - 4)^2 + z^2$
Elliptic cone.

e)

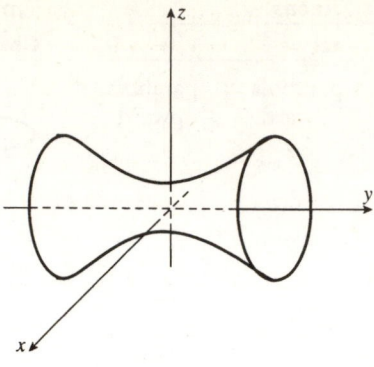

Hyperboloid of one sheet.

f)

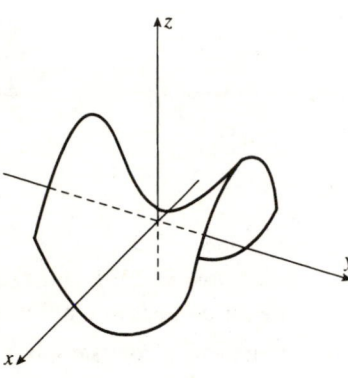

Hyperbolic paraboloid.

g)

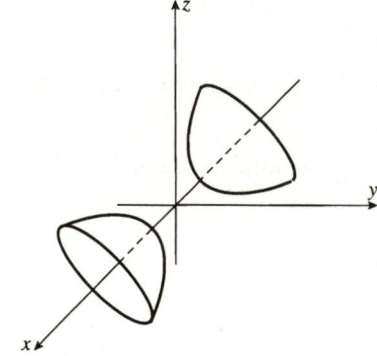

Hyperboloid of two sheets.

6) Sometimes, Always, or Never: The linear terms are important in determining the type of quadric surface.

Sometimes. If the equation contains both a first and a second power of a variable, the presence of the first power does not affect the *type* of surface it is.

Section 13.7 Cylindrical and Spherical Coordinates

This section describes two other coordinate systems for three-dimensional space. The first, cylindrical coordinates, is simply polar coordinates in the xy-plane with a z-axis added. The other, spherical coordinates, is the generalization of polar coordinates to three dimensions, where a distance and two angles are necessary to specify a point.

Concepts to Master

Cylindrical coordinates; Spherical coordinates; Conversion from one set of coordinates to another.

Summary and Focus Questions

Page 875 (ET Page 839)

The *cylindrical coordinates* (r, θ, z) of a point P in space are polar coordinates (r, θ) in the xy-plane and the usual z coordinate. They are called cylindrical because $r = a$ is a circular cylinder about the z-axis.

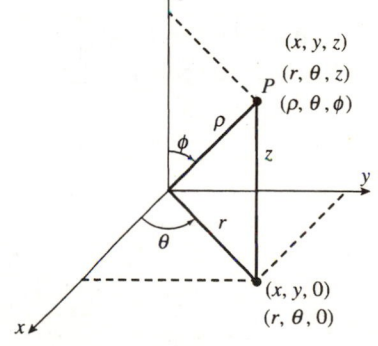

The *spherical coordinates* (ρ, θ, ϕ) of a point P in space are defined as:

ρ = distance from P to the origin.

θ = same as cylindrical coordinates.

ϕ = angle between the positive z-axis and the line from P to the origin.

They are called spherical coordinates because the equation $\rho = a$ is a sphere about the origin with radius a.

Thus we have three different coordinate systems for three-dimensional space. The following table shows how to convert coordinates for one system to the others.

To

From	Rectangular (x, y, z)	Cylindrical (r, θ, z)	Spherical (ρ, θ, ϕ)
Rectangular (x, y, z)		$r^2 = x^2 + y^2$ $\tan\theta = \frac{y}{x}$ $z = z$	$\rho^2 = x^2 + y^2 + z^2$ $\tan\theta = \frac{y}{x}$ $\cos\phi = \frac{z}{\rho}$
Cylindrical (r, θ, z)	$x = r\cos\theta$ $y = r\sin\theta$ $z = z$		$\rho^2 = r^2 + z^2$ $\theta = \theta$ $\cos\phi = \frac{z}{\rho}$
Spherical (ρ, θ, ϕ)	$x = \rho\sin\phi\cos\theta$ $y = \rho\sin\phi\sin\theta$ $z = \rho\cos\phi$	$r = \rho\sin\phi$ $\theta = \theta$ $z = \rho\cos\phi$	

1) Find the indicated coordinates of each point:

a) Spherical: $\left(8, \frac{\pi}{3}, \frac{\pi}{6}\right)$.
Rectangular: ______.

$(2, 2\sqrt{3}, 4\sqrt{3})$.
$x = 8\sin\frac{\pi}{6}\cos\frac{\pi}{3} = 8 \cdot \frac{1}{2} \cdot \frac{1}{2} = 2.$
$y = 8\sin\frac{\pi}{6}\sin\frac{\pi}{3} = 2\sqrt{3}.$
$z = 8\cos\frac{\pi}{6} = 8\frac{\sqrt{3}}{2} = 4\sqrt{3}.$

b) Cylindrical: $\left(\sqrt{3}, \frac{\pi}{6}, 1\right)$.
Spherical: ______.

$\left(2, \frac{\pi}{6}, \frac{\pi}{3}\right)$.
$\rho^2 = \sqrt{3}^2 + 1^2 = 4$, so $\rho = 2$.
$\theta = \frac{\pi}{6}$.
$\cos\phi = \frac{1}{2}$, so $\phi = \frac{\pi}{3}$.

c) Rectangular: $(2, 2\sqrt{3}, 6)$.
Cylindrical: ______.

$\left(4, \frac{\pi}{3}, 6\right)$.
$r^2 = 2^2 + (2\sqrt{3})^2 = 16, r = 4.$
$\tan\theta = \frac{2\sqrt{3}}{2} = \sqrt{3}, \theta = \frac{\pi}{3}.$
$z = 6.$

d) Spherical: $\left(10, \frac{\pi}{6}, \frac{\pi}{3}\right)$.
Cylindrical: _____.

$\left(5\sqrt{3}, \frac{\pi}{6}, 5\right)$.
$r = 10 \sin \frac{\pi}{3} = 10\left(\frac{\sqrt{3}}{2}\right) = 5\sqrt{3}$.
$\theta = \frac{\pi}{6}$.
$z = 10 \cos \frac{\pi}{3} = 10\left(\frac{1}{2}\right) = 5$.

e) Cylindrical: $\left(4, \frac{\pi}{4}, 3\right)$.
Rectangular: _____.

$(2\sqrt{2}, 2\sqrt{2}, 3)$.
$x = 4 \cos \frac{\pi}{4} = 4\frac{\sqrt{2}}{2} = 2\sqrt{2}$.
$y = 4 \sin \frac{\pi}{4} = 4\frac{\sqrt{2}}{2} = 2\sqrt{2}$.
$z = 3$.

f) Rectangular: $(1, 2, 2)$.
Spherical: _____.

$(3, 1.107, 0.841)$
$\rho^2 = 1^2 + 2^2 + 2^2 = 9, \rho = 3$.
$\tan \theta = \frac{2}{1}, \theta = \tan^{-1} 2 \approx 1.107$.
$\cos \phi = \frac{2}{3}, \phi = \cos^{-1} \frac{2}{3} \approx 0.841$.

2) Identify the surface with equation:

a) $\rho = \cos \phi$ in spherical coordinates.

Switch to rectangular coordinates.
$\rho = \cos \phi$
$\rho^2 = \rho \cos \phi$
$x^2 + y^2 + z^2 = z$
$x^2 + y^2 + \left(z - \frac{1}{2}\right)^2 = \frac{1}{4}$.
This is a sphere with radius $\frac{1}{2}$ and center $\left(0, 0, \frac{1}{2}\right)$.

b) $z^2 = r^2$ in cylindrical coordinates.

Switch to rectangular: $z^2 = x^2 + y^2$.
This is an elliptic cone.

Technology Plus for Chapter 13

1) Many calculators and CASs have a built-in vector cross product function (for example, "crossP" on the TI-92). Find $\mathbf{a} \times \mathbf{b}$ for

   a) $\mathbf{a} = \langle 3, 7, 2 \rangle$
   $\mathbf{b} = \langle 1, 2, 1 \rangle$

   b) $\mathbf{a} = \langle -3.51, 2.62, 8.01 \rangle$
   $\mathbf{b} = \langle 4.41, -1.02, 6.32 \rangle$

2) Use a calculator or computer with three-dimensional graphing capabilities to graph

$z = x^2 + 2xy,$
$-2 \leq x \leq 2, -2 \leq y \leq 2.$

Try different viewpoints.

$\mathbf{a} \times \mathbf{b} = \langle 3, -1, -1 \rangle$

$\mathbf{a} \times \mathbf{b} = \langle 24.73, 57.51, -7.97 \rangle$

With viewing angles of 15° for the x-axis and 70° for the z-axis, the graph is:

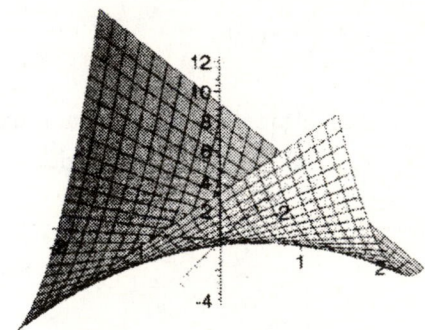

With viewing angles of 60° for the x-axis and 50° for the z-axis, the graph is:

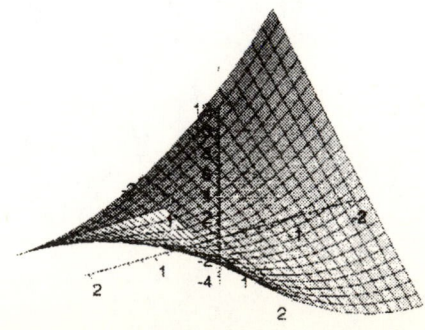

3) Sketch the graphs of these two lines:

L_1: $x = 1 - 3t$
$y = -3 - 7t$
$z = 2 + 4t$

L_2: $x = -5 - 4t$
$y = 6 + 2t$
$z = -1 + 3t$

Are the lines skew or do they intersect?

With viewing angles of $110°$ for x and $70°$ for z the graph is

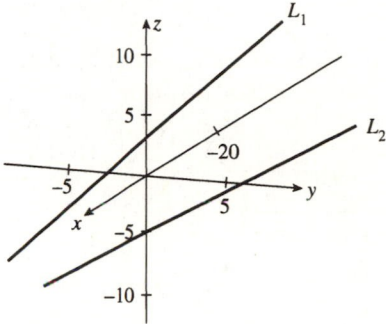

The lines are skew.

4) In cylindrical coordinates
$r = \theta, 0 \le \theta \le 2\pi$.

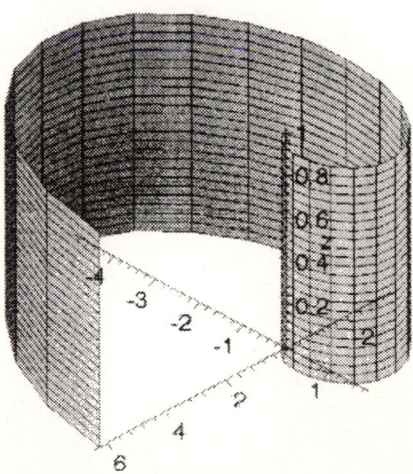

5) Graph $x^2 - y^2 + z^2 = 0$ for $-3 \le x \le 3$, $-3 \le y \le 3$, $-3 \le z \le 3$.

This is a cone which is not the graph of $z = f(x, y)$ for some function f. For many systems you will need to either plot it parametrically or use an implicit plotting command (In Maple V, for example, the command is

implicitplot3d ((x^2 − y^2 + z^2 = 0, x = −3..3, y = −3..3, z = −3..3);)

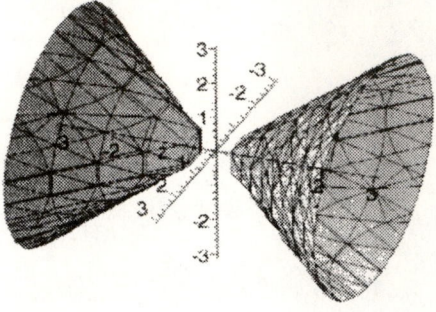

Chapter 14 — Vector Functions

Section 14.1 Vector Functions and Space Curves

This section extends the concepts of limits and continuity to vector functions. The "graphs" of vector functions are curves in two- or three-dimensional space. (In two dimensions, vector functions are another way to look at curves defined by a pair of parametric equations.)

Concepts to Master

Vector functions; Limits; Continuity; Curves in space

Summary and Focus Questions

Page 885 (ET Page 849)*

A *vector function* (or vector valued function) is a function having a set of real numbers as its domain and a set of vectors as its range – to each real number is associated a vector.

Most of this section is concerned with three-dimensional vectors. A vector function $\mathbf{r}(t)$ may be written with real valued *component functions* f, g, and h:

$$\mathbf{r}(t) = f(t)\mathbf{i} + g(t)\mathbf{j} + h(t)\mathbf{k}$$

Limits are defined component-wise:

$$\lim_{t \to t_0} \mathbf{r}(t) = (\lim_{t \to t_0}(f(t))\mathbf{i} + (\lim_{t \to t_0}(g(t))\mathbf{j} + (\lim_{t \to t_0}(h(t))\mathbf{k}.$$

$\mathbf{r}(t)$ is *continuous at a* means $\lim_{t \to a} \mathbf{r}(t) = \mathbf{r}(a)$.

The range of a vector function $\mathbf{r}(t)$ is a set of vectors. Those vectors, when thought of as points, trace out a *curve in space*. We say that the curve is defined parametrically by

$$x = f(t), y = g(t), z = h(t).$$

Visualizing the curve for a given vector function can be difficult. It may help to draw it in a box coordinate instead of simply with x, y, z axes or to determine what solid or surface the curve wraps itself around.

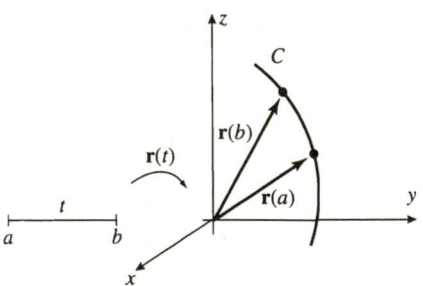

Example: Sketch the graph of the curve $\mathbf{r}(t)$ $= \langle \sin t \cos t, 1 + \sin 8t \rangle$ for $0 \leq t \leq 2\pi$. Let $x = \sin t$, $y = \cos t$, and $z = 1 + \sin 8t$. It helps to observe that z oscillates between 0 and 2 eight times and that $x^2 + y^2 = 1$. Thus the curve will be on the cylinder $x^2 + y^2 = 1$.

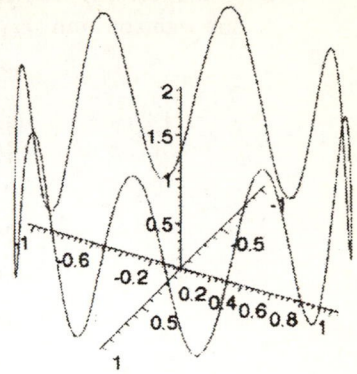

1) What is the domain of
$\mathbf{r}(t) = 2t\mathbf{i} - \frac{1}{t}\mathbf{j} + \sqrt{t+1}\ \mathbf{k}$?

$2t$ is defined for all t. $\frac{1}{t}$ is defined for all $t \neq 0$. $\sqrt{t+1}$ is defined for $t \geq -1$. The domain of $\mathbf{r}(t)$ is $[-1, 0) \cup (0, \infty)$.

2) For $\mathbf{r}(t) = \langle \sin t, t^2 + 4, e^t \rangle$

$\lim\limits_{t \to 0} \mathbf{r}(t) = $ _____.

$\langle \sin 0, 0^2 + 4, e^0 \rangle = \langle 0, 4, 1 \rangle$.

3) Sketch a graph of
$\mathbf{r}(t) = \langle 1 + \sin t \rangle \mathbf{i} + \langle 1 + \sin t \rangle \mathbf{j} + t\mathbf{k}$.

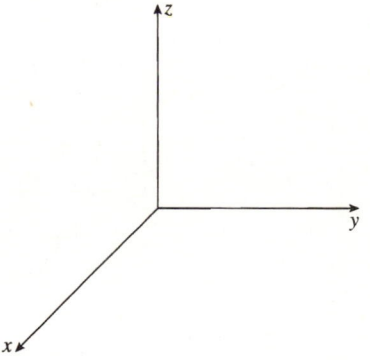

We note that $x = y$ so the curve is in the plane $x = y$ parallel to the z-axis.

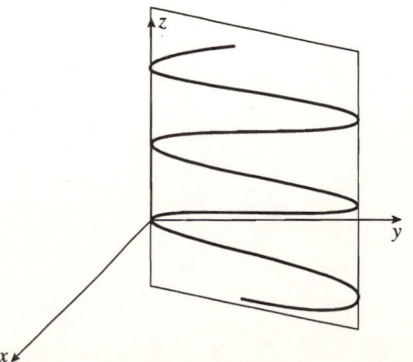

4) Find a vector function whose graph is the line segment from $(1, 2, 1)$ to $(4, 6, 5)$.

Let $x = 1 + (4 - 1)t = 1 + 3t$
$y = 2 + (6 - 2)t = 2 + 4t$
$z = 1 + (5 - 1)t = 1 + 4t.$

Then $\mathbf{r}(t) = \langle 1 + 3t, 2 + 4t, 1 + 4t \rangle$ for $0 \le t \le 1$ has a graph that is the line segment from $(1, 2, 1)$ to $(4, 6, 5)$.

Section 14.2 Derivatives and Integrals of Vector Functions

This section extends the concepts of derivatives and integrals to vector functions. As you might expect, differentiation and integration is performed component-wise.

Concepts to Master

A. Derivatives of vector functions; Tangent vector; Properties of derivatives
B. Smooth curves; Piecewise smooth curves
C. Integral of vector functions

Summary and Focus Questions

Page 893 (ET Page 857)

A. The *derivative* of a vector function $\mathbf{r}(t)$ is defined to be

$$\mathbf{r}'(t) = \lim_{h \to 0} \frac{\mathbf{r}(t+h) - \mathbf{r}(t)}{h}.$$

$\mathbf{r}'(t)$ is the *tangent vector* to the curve defined by the function $\mathbf{r}(t)$.

$\mathbf{T}(t) = \dfrac{\mathbf{r}'(t)}{|\mathbf{r}'(t)|}$ is the *unit tangent vector.*

If $\mathbf{r}(t) = f(t)\mathbf{i} + g(t)\mathbf{j} + h(t)\mathbf{k}$, then $\mathbf{r}'(t) = f'(t)\mathbf{i} + g'(t)\mathbf{j} + h'(t)\mathbf{k}$.

For differentiable vector functions $\mathbf{r}_1(t)$ and $\mathbf{r}_2(t)$ and scalar function $a(t)$:

$$[\mathbf{r}_1(t) + \mathbf{r}_2(t)]' = \mathbf{r}_1'(t) + \mathbf{r}_2'(t)$$
$$[c\,\mathbf{r}_1(t)]' = c[\mathbf{r}_1'(t)] \qquad (c, \text{ a constant})$$
$$[a(t)\mathbf{r}_1(t)]' = a'(t)\mathbf{r}_1(t) + a(t)\mathbf{r}_1'(t)$$
$$[\mathbf{r}_1(t) \cdot \mathbf{r}_2(t)]' = \mathbf{r}_1'(t) \cdot \mathbf{r}_2(t) + \mathbf{r}_1(t) \cdot \mathbf{r}_2'(t)$$
$$[\mathbf{r}_1(t) \times \mathbf{r}_2(t)]' = \mathbf{r}_1'(t) \times \mathbf{r}_2(t) + \mathbf{r}_1(t) \times \mathbf{r}_2'(t)$$
$$[\mathbf{r}_1(a(t))]' = a'(t)\mathbf{r}_1'(a(t)) \qquad (\text{Chain Rule})$$

1) Find $\mathbf{r}'(t)$ for $\mathbf{r}(t) = \langle e^{2t}, \tan t, t^3 \rangle$.

$$\mathbf{r}'(t) = \langle 2e^{2t}, \sec^2 t, 3t^2 \rangle.$$

2) Find the unit tangent vector to the curve $\mathbf{r}(t) = t^2\mathbf{i} - t^3\mathbf{j} + t^4\mathbf{k}$ at $t = 1$.

$$\mathbf{r}'(t) = 2t\mathbf{i} - 3t^2\mathbf{j} + 4t^3\mathbf{k}$$
$$\mathbf{r}'(1) = 2\mathbf{i} - 3\mathbf{j} + 4\mathbf{k}$$
$$\mathbf{T}(1) = \frac{2\mathbf{i} - 3\mathbf{j} + 4\mathbf{k}}{\sqrt{2^2 + (-3)^2 + 4^2}}$$
$$= \frac{2}{\sqrt{29}}\mathbf{i} - \frac{3}{\sqrt{29}}\mathbf{j} + \frac{4}{\sqrt{29}}\mathbf{k}.$$

3) Find $\mathbf{r}''(t)$ for $\mathbf{r}(t) = \langle \cos t, t^3 \rangle$.

$$\mathbf{r}'(t) = \langle -\sin t, 3t^2 \rangle$$
$$\mathbf{r}''(t) = \langle -\cos t, 6t \rangle$$

4) Find $[\mathbf{f}(t) \cdot \mathbf{g}(t)]'$ for
$$\mathbf{f}(t) = t^2\mathbf{i} + t^3\mathbf{j}$$
$$\mathbf{g}(t) = 7t\mathbf{i} + 3t^2\mathbf{j}$$

$$
\begin{aligned}
[\mathbf{f}(t) \cdot \mathbf{g}(t)]' &= \mathbf{f}'(t) \cdot \mathbf{g}(t) + \mathbf{f}(t) \cdot \mathbf{g}'(t) \\
&= [2t\mathbf{i} + 3t^2\mathbf{j}] \cdot [7t\mathbf{i} + 3t^2\mathbf{j}] \\
&\quad + [t^2\mathbf{i} + t^3\mathbf{j}] \cdot [7\mathbf{i} + 6t\mathbf{j}] \\
&= 14t^2 + 9t^4 + 7t^2 + 6t^4 \\
&= 21t^2 + 15t^4.
\end{aligned}
$$

B. A curve C given by a vector function $\mathbf{r}(t)$ for $a \le t \le b$ is *smooth* if $\mathbf{r}'(t)$ is continuous and $\mathbf{r}'(t) \ne \mathbf{0}$ for all $a < x < b$. C is *piecewise smooth* if C is made up of a finite number of smooth pieces.

Page 894 (ET Page 858)

Example: The curve determined by $\mathbf{r}(t) = \langle \cos t, \sin t, t^2 \rangle$ is smooth for any interval $[a, b]$ because $\mathbf{r}'(t) = \langle -\sin t, \cos t, 2t \rangle$ is never the zero vector ($|\mathbf{r}'(t)| = \sqrt{(-\sin t)^2 + (\cos t)^2 + (2t)^2} = \sqrt{1 + 4t^2}$ is never zero).

5) For each curve C determine whether it is smooth.

   a) $\mathbf{r}(t) = \langle e^t, t^2, t^3 \rangle$, $t \in [0, 1]$.

$\mathbf{r}'(t) = \langle e^t, 2t, 3t^2 \rangle$. Since the x-coordinate of $\mathbf{r}'(t)$ is never 0, $\mathbf{r}'(t) \ne \mathbf{0}$. Therefore, C is smooth everywhere and, in particular, for $t \in [0, 1]$.

   b) $\mathbf{r}(t) = \langle t \ln t - t, t^2 - 2t, e^t - t \rangle$ for $\frac{1}{2} \le t \le 2$.

$\mathbf{r}'(t) = \langle \ln t, 2t - 2, e^t - 1 \rangle$. At $t = 1$, $\mathbf{r}'(1) = \langle 0, 0, 0 \rangle = \mathbf{0}$. Therefore, C is not smooth for $\frac{1}{2} \le t \le 2$. C is, however, piecewise smooth. The two smooth pieces correspond to $\frac{1}{2} \le t \le 1$ and $1 \le t \le 2$.

Page
881
(ET Page
847)

C. If $\mathbf{r}(t) = f(t)\mathbf{i} + g(t)\mathbf{j} + h(t)\mathbf{k}$ is continuous on $[a, b]$,

$$\int_a^b \mathbf{r}(t)dt = \left(\int_a^b f(t)dt\right)\mathbf{i} + \left(\int_a^b g(t)dt\right)\mathbf{j} + \left(\int_a^b h(t)dt\right)\mathbf{k},$$

or, written in vector form:

$$\int_a^b \mathbf{r}(t)dt = \mathbf{R}(b) - \mathbf{R}(a), \text{ where } \mathbf{R}'(t) = \mathbf{r}(t).$$

6) Evaluate $\int_1^3 \left(\mathbf{i} + 2t\mathbf{j} + \frac{1}{t}\mathbf{k}\right)dt.$

The integral equals

$$\left(\int_1^3 dt\right)\mathbf{i} + \left(\int_1^3 2t\, dt\right)\mathbf{j} + \left(\int_1^3 \frac{1}{t}\, dt\right)\mathbf{k}$$

$$= 2\mathbf{i} + 8\mathbf{j} + (\ln 3)\mathbf{k}.$$

7) Find $\mathbf{r}(t)$ if $\mathbf{r}'(t) = e^{2t}\mathbf{i} - \mathbf{j} + t^2\mathbf{k}$ and
$\mathbf{r}(0) = \mathbf{i} - \mathbf{j}$.

$$\int(e^{2t}\mathbf{i} - \mathbf{j} + t^2\mathbf{k})dt = \frac{e^{2t}}{2}\mathbf{i} - t\mathbf{j} + \frac{t^3}{3}\mathbf{k} + \mathbf{C}$$

At $t = 0$, $\mathbf{r}(0) = \mathbf{i} - \mathbf{j}$. Thus $\frac{1}{2}\mathbf{i} + \mathbf{C} = \mathbf{i} - \mathbf{j}$

$$\mathbf{C} = \frac{1}{2}\mathbf{i} - \mathbf{j}.$$
$$\mathbf{r}(t) = \frac{e^{2t} + 1}{2}\mathbf{i} - (t + 1)\mathbf{j} + \frac{t^3}{3}\mathbf{k}.$$

Section 14.3   Arc Length and Curvature

This section shows you how to determine the arc length of a smooth curve. Not surprisingly, the process is the same as that for two-dimensional curves defined parametrically in chapter 11. This section also contains several concepts for characterizing the shape of a curve; for example, curvature – a measure of how fast the curve is bending at a given point.

Concepts to Master

A. Length of an arc in space; Arc length function; Reparametrization

B. Curvature

C. Tangent vector; Unit normal vector; Normal plane

Summary and Focus Questions

Page 898 (ET Page 862)

A. If $a \le t \le b$ and $\mathbf{r}(t) = \langle f(t), g(t), h(t) \rangle$ is smooth, the *length of the curve* is

$$L = \int_a^b |\mathbf{r}'(t)| dt = \int_a^b \sqrt{[f'(t)]^2 + [g'(t)]^2 + [h'(t)]^2} \, dt .$$

A curve may be given by more than one vector function – these are different *parametrizations* of the same curve.

Example: The quarter circle C may be parametrized two ways:

$$\mathbf{r}(t) = \langle \sin t, \cos t \rangle, 0 \le t \le \frac{\pi}{2}$$
$$\mathbf{q}(t) = \langle t, \sqrt{1 - t^2} \rangle, 0 \le t \le 1$$

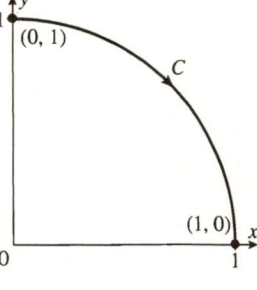

Both traverse the curve C from $(0, 1)$ to $(1, 0)$ but at different speeds along the way. Regardless, the arc length is the same:

$$\int_0^{\frac{\pi}{2}} |\mathbf{r}'(t)| dt = \int_0^{\frac{\pi}{2}} \sqrt{\cos^2 t + \sin^2 t} \, dt = \int_0^{\frac{\pi}{2}} 1 dt = \frac{\pi}{2}.$$

$$\int_0^1 |\mathbf{q}'(t)| dt = \int_0^1 \sqrt{1 + \left(\frac{-t}{\sqrt{1 - t^2}}\right)^2} \, dt = \int_0^1 \frac{1}{\sqrt{1 - t^2}} dt = \sin^{-1} t \Big|_0^1 = \frac{\pi}{2}.$$

The *arc length function* is $s(t) = \int_a^t |\mathbf{r}'(u)| du$ and therefore $\frac{ds}{dt} = |\mathbf{r}'(t)|$.

If we can solve for t in terms of s (so that we can write $t = t(s)$), we can reparametrize the curve – that is, write it in the form $\mathbf{r} = \mathbf{r}(t(s))$. This particular reparametrization represents moving a point from the beginning of the curve to the end with uniform speed (with respect to distance travelled along the curve.)

1) Find the length of the arc given by
$r(t) = \ln t\,\mathbf{i} - t^2\mathbf{j} + 2t\mathbf{k}$ for $t \in [1, 4]$.

$r'(t) = \frac{1}{t}\mathbf{i} - 2t\mathbf{j} + 2\mathbf{k}.$

$|r'(t)| = \sqrt{\frac{1}{t^2} + 4^2 + 4} = \sqrt{\frac{1 + 4t^2 + 4t^4}{t^2}}$

$= \sqrt{\frac{(1 + 2t^2)^2}{t^2}} = \frac{1 + 2t^2}{t} = \frac{1}{t} + 2t.$

The length of the arc is

$$L = \int_1^4 \left(\frac{1}{t} + 2t\right)dt = \left.(\ln t + t^2)\right]_1^4$$

$$= 15 + \ln 4.$$

2) a) Find the arc length function for the curve C given by
$r(t) = (1 - t^2)\mathbf{i} + t^2\mathbf{j} + \sqrt{2}t^2\mathbf{k}$
for $0 \le t \le 4$.

$r'(t) = -2t\mathbf{i} + 2t\mathbf{j} + 2\sqrt{2}t\mathbf{k}.$

$|r'(t)| = \sqrt{(-2t)^2 + (2t)^2 + (2\sqrt{2}t)^2} = 4t$

$s(t) = \int_0^t |r'(u)|\,du = \int_0^t 4u\,du = 2t^2.$

b) Reparametrize the curve C using the arc length function.

From part a), $s = 2t^2$, $t^2 = \frac{s}{2}$ and $t = \sqrt{\frac{s}{2}}$.

Thus the reparametrization is

$r(t(s)) = \left(1 - \frac{s}{2}\right)\mathbf{i} + \frac{s}{2}\mathbf{j} + \frac{s}{\sqrt{2}}\mathbf{k}.$

Page 900 (ET Page 864)

B. The *curvature* κ for C parameterized by a twice differentiable $r(t)$ is

$$\kappa = \left|\frac{d\mathbf{T}}{ds}\right| = \left|\frac{\mathbf{T}'(t)}{\mathbf{r}'(t)}\right| = \frac{|\mathbf{r}'(t) \times \mathbf{r}''(t)|}{|\mathbf{r}'(t)|^3}.$$

In the special case of $y = f(x)$ in the plane this becomes

$$\kappa = \frac{|y''|}{[1 + (y')^2]^{3/2}}.$$

The curvature measures how quickly a curve C changes direction.

3) Which curve would have greater curvature at $t = 0$?

C_1: $x = t$ $\qquad\qquad$ C_2: $x = t$
$\quad y = t^2$ $\qquad\qquad\qquad\quad y = t^4$

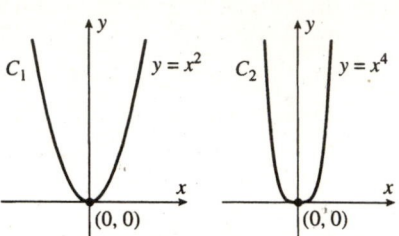

At $(0, 0)$, C_1 has a greater curvature. (Curve C_2 is flatter near $(0, 0)$.)

4) Find the curvature at $t = 1$ for the curve
$\mathbf{r}(t) = t^2\mathbf{i} + \frac{2}{3}t^3\mathbf{j} + 2t\mathbf{k}$.

$\mathbf{r}'(t) = 2t\mathbf{i} + 2t^2\mathbf{j} + 2\mathbf{k}$.
$|\mathbf{r}'(t)| = \sqrt{4t^2 + 4t^4 + 4} = 2\sqrt{t^2 + t^4 + 1}$
$\mathbf{r}''(t) = 2\mathbf{i} + 4t\mathbf{j}$.

$$\mathbf{r}'(t) \times \mathbf{r}''(t) = \begin{vmatrix} \mathbf{i} & \mathbf{j} & \mathbf{k} \\ 2t & 2t^2 & 2 \\ 2 & 4t & 0 \end{vmatrix}$$

$$= -8t\mathbf{i} + 4\mathbf{j} + 4t^2\mathbf{k}$$

Thus $|\mathbf{r}' \times \mathbf{r}''| = \sqrt{64t^2 + 16 + 16t^4}$
$$= 4\sqrt{4t^2 + 1 + t^4}.$$

Therefore $\kappa = \dfrac{|\mathbf{r}' \times \mathbf{r}''|}{|\mathbf{r}'|^3} = \dfrac{4\sqrt{4t^2 + 1 + t^4}}{(2\sqrt{t^2 + t^4 + 1})^3}$

$$= \dfrac{4\sqrt{4t^2 + 1 + t^4}}{8(t^2 + t^4 + 1)^{3/2}}$$

At $t = 1$, $\kappa = \dfrac{\sqrt{4 + 1 + 1}}{2(1 + 1 + 1)^{3/2}} = \dfrac{\sqrt{6}}{2(3)^{3/2}}$

$$= \dfrac{\sqrt{6}}{2(3)\sqrt{3}} = \dfrac{\sqrt{2}}{6}.$$

5) Find the curvature of $y = x^2$ at $x = \sqrt{6}$.

For $y = x^2$, $y' = 2x$, $y'' = 2$ and

$$\kappa = \dfrac{|2|}{(1 + (2x)^2)^{3/2}}.$$

At $x = \sqrt{6}$, $\kappa = \dfrac{2}{(25)^{3/2}} = \dfrac{2}{125}$.

Page
903
(ET Page
867)

C. For a smooth curve C given by the function $\mathbf{r}(t)$, the *unit tangent vector* $\mathbf{T}$ is $\mathbf{T}(t) = \frac{\mathbf{r}'(t)}{|\mathbf{r}'(t)|}$ and the principal *unit normal vector* $\mathbf{N}$ is $\mathbf{N}(t) = \frac{\mathbf{T}'}{|\mathbf{T}'|}$.

$\mathbf{N}$ is orthogonal to $\mathbf{T}$ and always points "inward" on the curve C. The binormal vector $\mathbf{B}$ is $\mathbf{B}(t) = \mathbf{T}(t) \times \mathbf{N}(t)$.

The plane formed by $\mathbf{N}$ and $\mathbf{B}$ is the *normal plane*; all vectors in the normal plane are orthogonal to $\mathbf{T}$. Thus the curve C pierces the normal plane at a right angle.

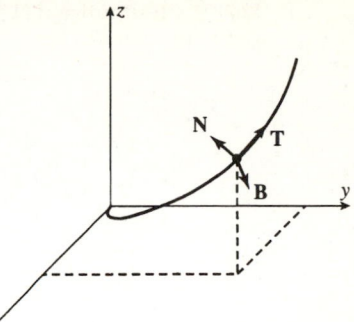

The plane formed by $\mathbf{T}$ and $\mathbf{N}$ is called the *osculating plane*; it is the plane that best approximates the direction of the curve. We may draw a circle in that plane with radius $\frac{1}{\kappa}$ and through the point on the curve C. This *osculating circle* has the same tangent vector, normal vector and curvature as C (and, therefore, is a good approximation to the curve).

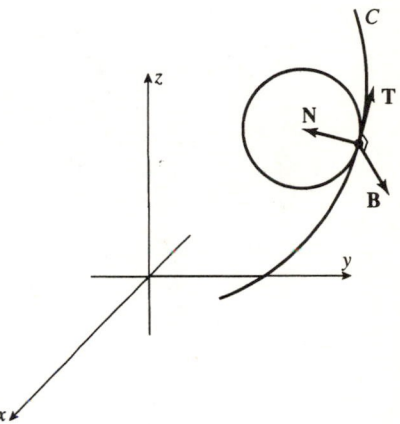

6) Find the principal unit normal vector for $\mathbf{r}(t) = \sin t\mathbf{i} + 2t\mathbf{j} - \cos t\mathbf{k}$ at $t = \frac{\pi}{4}$.

$\mathbf{r}'(t) = \cos t\mathbf{i} + 2\mathbf{j} + \sin t\mathbf{k}$.
$|\mathbf{r}'(t)| = \sqrt{\cos^2 t + 4 + \sin^2 t} = \sqrt{5}$.
Thus $\mathbf{T} = \frac{1}{\sqrt{5}}(\cos t\mathbf{i} + 2\mathbf{j} + \sin t\mathbf{k})$.
$\mathbf{T}' = \frac{1}{\sqrt{5}}(-\sin t\mathbf{i} + \cos t\mathbf{k})$.
$|\mathbf{T}'| = \frac{1}{\sqrt{5}}\sqrt{\sin^2 t + \cos^2 t} = \frac{1}{\sqrt{5}}$.
Thus $\mathbf{N} = \frac{\mathbf{T}'}{|\mathbf{T}'|} = -\sin t\mathbf{i} + \cos t\mathbf{k}$.
At $t = \frac{\pi}{4}$, $\mathbf{N} = -\frac{\mathbf{i}}{\sqrt{2}} + \frac{\mathbf{k}}{\sqrt{2}}$.

7) For $\mathbf{r}(t) = 2t\mathbf{i} - \frac{t^2}{2}\mathbf{j} + t\mathbf{k}$ find
   a) the unit tangent vector $\mathbf{T}$ at $t = 2$

$\mathbf{r}'(t) = 2\mathbf{i} - t\mathbf{j} + \mathbf{k}$
$|\mathbf{r}'(t)| = \sqrt{4 + t^2 + 1} = \sqrt{t^2 + 5}$.
$\mathbf{T}(t) = \frac{2}{\sqrt{t^2 + 5}}\mathbf{i} - \frac{t}{\sqrt{t^2 + 5}}\mathbf{j} + \frac{1}{\sqrt{t^2 + 5}}\mathbf{k}$.
$\mathbf{T}(2) = \frac{2}{3}\mathbf{i} - \frac{2}{3}\mathbf{j} + \frac{1}{3}\mathbf{k}$.

b) the unit normal **N** at $t = 2$

$$T'(t) = \frac{-2t}{(t^2 + 5)^{3/2}}\mathbf{i} - \frac{5}{(t^2 + 5)^{3/2}}\mathbf{j} - \frac{t}{(t^2 + 5)^{3/2}}\mathbf{k}.$$

$$T'(2) = -\frac{4}{27}\mathbf{i} - \frac{5}{27}\mathbf{j} - \frac{2}{27}\mathbf{k}.$$

$$|T'(2)| = \sqrt{\frac{16 + 25 + 4}{27^2}} = \frac{\sqrt{5}}{9}.$$

Therefore, $\mathbf{N}(2) = \dfrac{T'(2)}{|T'(2)|}$

$$= \frac{9}{\sqrt{5}}\left(-\frac{4}{27}\mathbf{i} - \frac{5}{27}\mathbf{j} - \frac{2}{27}\mathbf{k}\right)$$

$$= -\frac{4}{3\sqrt{5}}\mathbf{i} - \frac{5}{3\sqrt{5}}\mathbf{j} - \frac{2}{3\sqrt{5}}\mathbf{k}.$$

c) the binormal **B** at $t = 2$

$$\mathbf{B}(2) = \mathbf{T}(2) \times \mathbf{N}(2) = \begin{vmatrix} \mathbf{i} & \mathbf{j} & \mathbf{k} \\ \frac{2}{3} & -\frac{2}{3} & \frac{1}{3} \\ \frac{-4}{3\sqrt{5}} & \frac{-5}{3\sqrt{5}} & \frac{-2}{3\sqrt{5}} \end{vmatrix}$$

$$= \frac{1}{\sqrt{5}}\mathbf{i} - 0\mathbf{j} + \frac{2}{\sqrt{5}}\mathbf{k}.$$

d) the normal plane at $t = 2$

The normal plane contains vectors orthogonal to $\mathbf{T}(2) = \left\langle\frac{2}{3}, -\frac{2}{3}, \frac{1}{3}\right\rangle$.

$\mathbf{r}(2) = \langle 4, -2, 2\rangle$.

$\mathbf{T}(2) \cdot \mathbf{r}(2) = 4\left(\frac{2}{3}\right) - 2\left(-\frac{2}{3}\right) + 2\left(\frac{1}{3}\right) = \frac{14}{3}.$

The equation of the plane is

$\frac{2}{3}(x - 4) - \frac{2}{3}(y + 2) + \frac{1}{3}(z - 2) = \frac{14}{3},$

or $2x - 2y + z = 28.$

e) the osculating plane at $t = 2$

This plane contains vectors orthogonal to $\mathbf{B}(2) = \left\langle\frac{1}{\sqrt{5}}, 0, \frac{2}{\sqrt{5}}\right\rangle$.

$\mathbf{B}(2) \cdot \mathbf{r}(2) = 4\left(\frac{1}{\sqrt{5}}\right) - 2(0) + 2\left(\frac{2}{\sqrt{5}}\right)$

$$= \frac{8}{\sqrt{5}}.$$

The equation of the plane is

$$\frac{1}{\sqrt{5}}(x - 4) + 0(y + 2) + \frac{2}{\sqrt{5}}(z - 2) = \frac{8}{\sqrt{5}},$$

or $x + 2z = 16.$

Section 14.4   Motion in Space: Velocity and Acceleration

All previous concepts in this chapter, such as tangent vectors and curvature, are used in this section to describe motion of an object in three-dimensional space.

Concepts to Master

A. Velocity; Speed; Acceleration; Newton's Second Law of Motion

B. Tangent and normal components of acceleration

Summary and Focus Questions

Page 907 (ET Page 871)

A. A vector function $\mathbf{r}(t)$ may be thought of as the position of a particle in space at time t.

$\mathbf{v}(t) = \mathbf{r}'(t)$ is the *velocity* vector and $|\mathbf{r}'(t)|$ is the *speed*.

$\mathbf{a}(t) = \mathbf{r}''(t)$ is the *acceleration* vector.

The unit tangent normal $\mathbf{T}$ points in the direction the particle is moving and the unit normal vector $\mathbf{N}$ points in the direction orthogonal to the motion. Vector integrals may be used to determine velocity $\mathbf{v}(t) = \int \mathbf{a}(t)dt$ and position $\mathbf{r}(t) = \int \mathbf{v}(t)dt$.

Newton's Second Law of Motion is $\mathbf{F}(t) = m\mathbf{a}(t)$, where $\mathbf{F}(t)$ is the force acting on an object of mass m whose acceleration is $\mathbf{a}(t)$.

In the particular case of a projectile of mass m launched at an angle α with initial velocity $\mathbf{v}_0 = (\mathbf{v}_0 \cos \alpha)\mathbf{i} + (\mathbf{v}_0 \sin \alpha)\mathbf{j}$:

$$\mathbf{F} = -mg\mathbf{j} \quad (g = 9.8 \text{ m/s}^2)$$
$$\mathbf{a} = -g\mathbf{j}$$
$$\mathbf{v} = -gt\mathbf{j} + \mathbf{v}_0$$
$$\mathbf{r} = -\tfrac{1}{2}gt^2\mathbf{j} + t\mathbf{v}_0$$

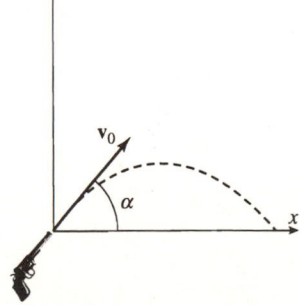

1) Find the velocity, speed, and acceleration of a particle whose position at time t is $\mathbf{r}(t) = \langle 3t^2, t^3 + 1, e^{-t} \rangle$.

$\mathbf{v}(t) = \mathbf{r}'(t) = \langle 6t, 3t^2, -e^{-t} \rangle$.

Speed is $|\mathbf{v}(t)| = \sqrt{36t^2 + 9t^4 + e^{-2t}}$.

$\mathbf{a}(t) = \langle 6, 6t, e^{-t} \rangle$.

2) Find the velocity and position at time t of a particle whose initial position is $\mathbf{i} + \mathbf{j}$, initial velocity is $\mathbf{j} + \mathbf{k}$, and acceleration is $\mathbf{a}(t) = 12t\mathbf{i} + 2\mathbf{k}$.

$\mathbf{v}(t) = \int \mathbf{a}(t)dt = 6t^2\mathbf{i} + 2t\mathbf{k} + \mathbf{C}$.

Initially $(t = 0)$ $\mathbf{v}_0 = \mathbf{j} + \mathbf{k}$:

$0\mathbf{i} + 0\mathbf{k} + \mathbf{C} = \mathbf{j} + \mathbf{k}$.

$\mathbf{v}(t) = 6t^2\mathbf{i} + 2t\mathbf{k} + (\mathbf{j} + \mathbf{k})$

$\quad = 6t^2\mathbf{i} + \mathbf{j} + (2t + 1)\mathbf{k}$.

$\mathbf{r}(t) = \int \mathbf{v}(t)dt = 2t^3\mathbf{i} + t\mathbf{j} + (t^2 + t)\mathbf{k} + \mathbf{D}$

Initially, $\mathbf{r}(0) = \mathbf{i} + \mathbf{j}$:

$0\mathbf{i} + 0\mathbf{j} + 0\mathbf{k} + \mathbf{D} = \mathbf{i} + \mathbf{j}$.

$\mathbf{r}(t) = 2t^3\mathbf{i} + t\mathbf{j} + (t^2 + t)\mathbf{k} + \mathbf{i} + \mathbf{j}$

$\quad = (2t^3 + 1)\mathbf{i} + (t + 1)\mathbf{j} + (t^2 + t)\mathbf{k}$.

3) What force is required in order for a 5 kg mass to be pushed in such a way that, at time t, its position is $\mathbf{r}(t) = 3t^2\mathbf{i} + t^4\mathbf{j} + 2t^3\mathbf{k}$?

$\mathbf{v}(t) = \mathbf{r}'(t) = 6t\mathbf{i} + 4t^3\mathbf{j} + 6t^2\mathbf{k}$.

$\mathbf{a}(t) = \mathbf{v}'(t) = 6\mathbf{i} + 12t^2\mathbf{j} + 12t\mathbf{k}$.

$\mathbf{F}(t) = 5\mathbf{a}(t) = 30\mathbf{i} + 60t^2\mathbf{j} + 60t\mathbf{k}$.

4) A projectile is fired with a velocity of 200 m/sec at an inclination of 30° from a point 10 meters above the ground. Find the vector function that describes the path of motion.

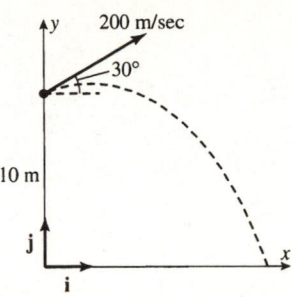

Choose a coordinate system so that from Newton's Second Law, $\mathbf{a}(t) = -g\mathbf{j}$.

Thus $\mathbf{v}(t) = \int -g\mathbf{j}\,dt = -gt\mathbf{j} + \mathbf{C}$.

At $t = 0$, $\mathbf{v}_0 = \mathbf{C}$. Since the velocity is 200 m/sec at a 30° inclination,

$\mathbf{C} = (200\cos 30°)\mathbf{i} + (200\sin 30°)\mathbf{j}$
$= 100\sqrt{3}\mathbf{i} + 100\mathbf{j}$.

Thus $\mathbf{v}(t) = 100\sqrt{3}\mathbf{i} + (100 - gt)\mathbf{j}$.

$\mathbf{r}(t) = \int \mathbf{v}(t)\,dt$
$= 100\sqrt{3}t\mathbf{i} + \left(100t - \frac{gt^2}{2}\right)\mathbf{j} + \mathbf{D}$.

At $t = 0$, $\mathbf{r}(0) = \mathbf{D}$. Since the projectile is 10 meters above the ground, $\mathbf{D} = 10\mathbf{j}$. Thus $\mathbf{r}(t) = 100\sqrt{3}t\mathbf{i} + \left(100t - \frac{gt^2}{2} + 10\right)\mathbf{j}$.

Page 911 (ET Page 875)

B. The acceleration vector **a** lies in the plane determined by the unit tangent vector **T** and the unit normal vector **N**. Thus **a** may be written as a combination of **T** and **N**:

$$\mathbf{a} = a_T\mathbf{T} + a_N\mathbf{N},$$

where the tangential component is $a_T = v'$ (v is the speed, $v = |\mathbf{v}|$) and the normal component is $a_N = \kappa v^2$ (κ is the curvature).

In terms of the position function $\mathbf{r}(t)$, $a_T = \dfrac{\mathbf{r}' \cdot \mathbf{r}''}{|\mathbf{r}'|}$ and $a_N = \dfrac{|\mathbf{r}' \cdot \mathbf{r}''|}{|\mathbf{r}'|}$.

5) Find the tangential and normal components of acceleration for $\mathbf{r}(t) = t^2\mathbf{i} + t^4\mathbf{j} + t^3\mathbf{k}$.

$\mathbf{r}'(t) = 2t\mathbf{i} + 4t^3\mathbf{j} + 3t^2\mathbf{k}.$

$|\mathbf{r}'(t)| = \sqrt{4t^2 + 16t^6 + 9t^4}$

$\qquad = t\sqrt{4 + 16t^4 + 9t^2}.$

$\mathbf{r}''(t) = 2\mathbf{i} + 12t^2\mathbf{j} + 6t\mathbf{k}.$

$\mathbf{r}' \cdot \mathbf{r}'' = 4t + 48t^5 + 18t^3.$

$\mathbf{r}' \times \mathbf{r}'' = \begin{vmatrix} \mathbf{i} & \mathbf{j} & \mathbf{k} \\ 2t & 4t^3 & 3t^2 \\ 2 & 12t^2 & 6t \end{vmatrix}$

$\qquad = -12t^4\mathbf{i} - 6t^2\mathbf{j} + 16t^3\mathbf{k}.$

$|\mathbf{r}' \times \mathbf{r}''| = \sqrt{144t^8 + 36t^4 + 256t^6}$

$\qquad = 2t^2\sqrt{36t^4 + 9 + 64t^2}.$

Thus $a_T = \dfrac{4t + 48t^5 + 18t^3}{t\sqrt{4 + 16t^4 + 9t^2}}$

$\qquad = \dfrac{4 + 48t^4 + 18t^2}{\sqrt{4 + 16t^4 + 9t^2}}$ and

$a_N = \dfrac{2t^2\sqrt{36t^4 + 9 + 64t^2}}{t\sqrt{4 + 16t^4 + 9t^2}}$

$\qquad = \dfrac{2t\sqrt{36t^4 + 9 + 64t^2}}{\sqrt{4 + 16t^4 + 9t^2}}.$

Technology Plus for Chapter 14

1) Use a computer algebra to graph $\mathbf{r}(t)$ where
$x = \cos 5t \cos 6t$
$y = \sin 5t \cos 6t$
$z = \sin 6t, \ 0 \le t \le 2\pi$.

On what surface does the curve lie?

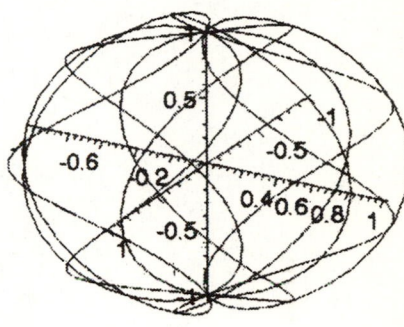

We note that $x^2 + y^2 + z^2 = 1$, so $\mathbf{r}(t)$ lies on the sphere with center $(0, 0, 0)$ and radius 1.

2) Find the parametric equations for the tangent line to the curve $\mathbf{r}(t) = \langle 4t, 2t^2, t^3 \rangle$ at $t = 1.5$. Graph the curve and unit tangent vector on the same screen.

$\mathbf{r}(t) = \langle 4t, 2t^2, t^3 \rangle$

$\mathbf{r}'(t) = \langle 4, 4t, 3t^2 \rangle$

At $t = 1.5$, $\mathbf{r}(1.5) = \langle 6, 4.5, 3.375 \rangle$ and $\mathbf{r}'(1.5) = \langle 4, 6, 6.75 \rangle$.

The tangent line is
$$x = 6 + 4t$$
$$y = 4.5 + 6t$$
$$z = 3.375 + 6.75t$$

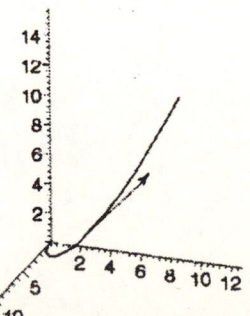

3) Graph $y = e^x$ and the osculating circle at $(0, 1)$.

For $y = e^x$, $\kappa = \dfrac{y''}{(1 + (y')^2)^{3/2}} = \dfrac{e^t}{(1 + (e^t)^2)^{3/2}}$

$= \dfrac{e^t}{(1 + e^{2t})^{3/2}}.$

At $t = 0$, $\kappa = \dfrac{1}{(1 + 1)^{3/2}} = \dfrac{1}{2\sqrt{2}}.$

Thus $\dfrac{1}{\kappa} = 2\sqrt{2}.$

The tangent at $(0, 1)$ has slope 1, so the normal has slope -1. The center of the osculating circle is on the normal:

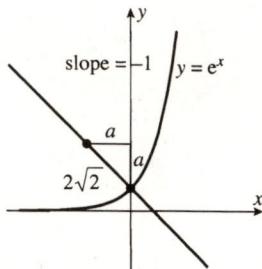

$a = 2$, so the center of the circle is $(-2, 3)$. The equation is $(x + 2)^2 + (y - 3)^2 = 8$. The graph of $y = e^x$ and the osculating circle are given below:

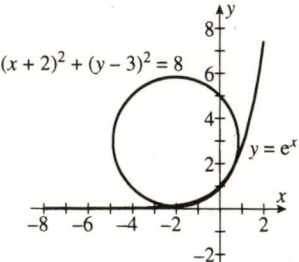

Chapter 15 — Partial Derivatives

"WHAT'S MOST DEPRESSING IS THE REALIZATION THAT EVERYTHING WE BELIEVE WILL BE DISPROVED IN A FEW YEARS."

Cartoons courtesy of Sidney Harris. Used by permission.

Section 15.1   Functions of Several Variables

In the last chapter we saw that the range of a function could be a multi-dimensional vector. In this section we study functions whose domains are multi-dimensional; that is, a function whose values are of the form $f(x, y)$ or $f(x, y, z)$. (Later in chapter 17 we will study functions with both domain and range multi-dimensional.) A function of two variables, $z = f(x, y)$, will have a graph in three dimensions (x, y, and z). Since these graphs are sometimes difficult to visualize, we introduce the concept of a "level curve" – a subset of the domain that correspond to a given functional value.

Concepts to Master

A. Functions of two variables; Domain; Range; Graphs

B. Level curves

C. Functions of more than two variables

Summary and Focus Questions

Page 923 (ET Page 887)*

A. A *function of two variables* assigns to each ordered pair (x, y) in $D \subset R^2$ a unique real number $f(x, y)$. D is the *domain* of f. The *range* of f is the set of all $f(x, y)$ values.

As is the case with a function of one variable, a function may be defined verbally, by a table of values, with a graph, or by an explicit formula.

There are three different ways to view the domain of f.
   1. f is a function of two independent variables, x and y, where $(x, y) \in D$.
   2. f is a function whose domain is all points (x, y) in D.
   3. f is a function whose domain is all vectors $\langle x, y \rangle$ in D.

For example, let $f(x, y) = x^2 + y^2$. Then for $x = 2, y = 1$
$f(2, 1) = 2^2 + 1^2 = 5$. The value 5 may be thought of as being associated with the values $x = 2$ and $y = 1$, or with the point $(2, 1)$, or with the vector $\langle 2, 1 \rangle$.

The graph of a function of two variables is a surface in three dimensional space. The point (x, y, z) is on the graph of f if and only if $z = f(x, y)$.

Our example, $f(x, y) = x^2 + y^2$ has a graph which is a circular paraboloid.

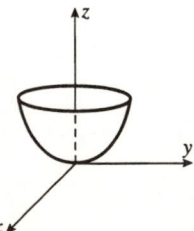

1) What is the domain of $f(x, y) = \dfrac{1}{5x - 10y}$?

$f(x, y)$ is defined for all (x, y) except when $5x - 10y = 0$ or $x = 2y$. The domain is $\{(x, y) | x \neq 2y\}$.

2) What are the domain and range of $f(x, y) = \dfrac{1}{x^2 + y^2}$?

$x^2 + y^2 = 0$ only for $(x, y) = (0, 0)$. The domain is $\{(x, y) | x \neq 0 \text{ and } y \neq 0\}$.
$x^2 + y^2 > 0$ for all $(x, y) \neq (0, 0)$. Thus the range is $(0, \infty)$.

3) Let $f(x, y)$ be defined by the table:

		y	
	4	5	6
1	8	7	−1
x 2	4	−10	7
3	3	−3	2

a) What is the domain?

The domain consists of the nine pairs: $(1, 4)$, $1, 5)$, $(1, 6)$, $(2, 4)$, $(2, 5)$, $(2, 6)$, $(3, 4)$, $(3, 5)$, $(3, 6)$.

b) What is the range?

$\{-10, -3, -1, 2, 3, 4, 7, 8\}$

c) What is $f(2, 5)$?

$f(2, 5) = -10$.

d) What is $f(4, 3)$?

$f(4, 3)$ does not exist.

4) Describe the graph of $f(x, y) = x^2$.

$z = x^2$ is a parabolic cylinder along the y-axis:

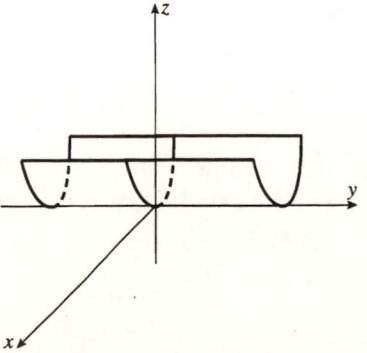

Page
928
(ET Page
892)

B. For a constant k, the *level curve determined by k* is $\{(x, y) \in D \mid f(x, y) = k\}$.

Each level curve is a subset of the domain of f. Level curves are not part of the graph of f. Level curves for various values of k may be drawn to obtain a visualization of the graph of $z = f(x, y)$, much in the same fashion as you see temperature regions on a weather map.

For $f(x, y) = x^2 + y^2$, each $k > 0$ produces the level curve $\{(x, y) \mid x^2 + y^2 = k\}$ which is a circle in the xy-plane.

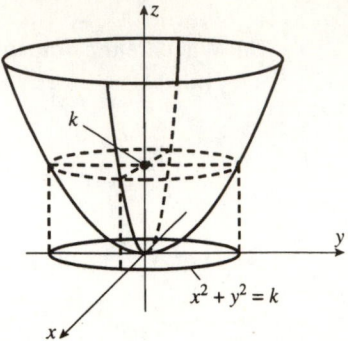

5) Sketch the level curves for $f(x, y) = x - y^2$ for $k = 0, 2, -2, 4$.

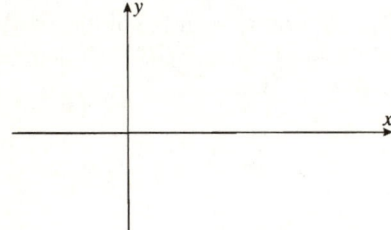

Each level curve has the form $x - y^2 = k$ which is a parabola opening to the right with vertex $(k, 0)$.

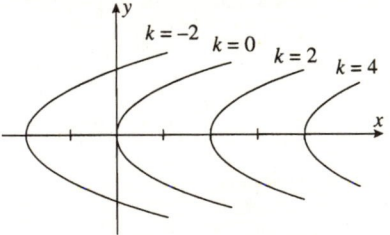

6) True or False:
Two different level curves can never intersect.

True, since every point in the domain has a unique functional value.

Page
932
(ET Page
896)

C. A function of three variables $f(x, y, z)$ has a graph in 4-dimensional space and level surfaces that are subsets of R^3. Functions of more than three variables and their level surfaces are defined similarly.

7) For $f(x, y, z) = x^2 - xy + z$ what is $f(2, 2, -1)$?

$$f(2, 2, -1) = 2^2 - (2)(2) + (-1) = -1.$$

8) Describe the level surfaces of $f(x, y, z) = x^2 + 4y^2 + 9z^2$.

For any $k \geq 0$, the level surfaces is all (x, y, z) such that $x^2 + 4y^2 + 9z^2 = k$, which is an ellipsoid with center $(0, 0, 0)$.

9) What is the domain of $f(x, y, z) = \dfrac{\sqrt{x}}{y+z}$?

$f(x, y, z)$ is defined for all $x \geq 0$ and whenever $y + z \neq 0$. The domain is $\{(x, y, z) | x \geq 0 \text{ and } y \neq -z\}$.

Section 15.2   Limits and Continuity

This section extends the concepts of limits and continuity to functions of two or more variables. Be careful to note that a function can be "one dimensionally" continuous along every line approaching a point and still fail to be continuous at that point.

Concepts to Master

A. Limits of functions of two or more variables; Limits along curves

B. Continuity of functions of two or more variables

Summary and Focus Questions

Page 938 (ET Page 902)

A. Let f be defined on a disk with center (a, b) except perhaps at (a, b).

$\lim\limits_{(x, y)\to(a, b)} f(x, y) = L$ means:

for all $\varepsilon > 0$ there exists $\delta > 0$ such that $\left|f(x, y) - L\right| < \varepsilon$ whenever $0 < \sqrt{(x - a)^2 + (y - b)^2} < \delta$.

Using $\mathbf{x} = \langle x, y\rangle$ and $\mathbf{a} = \langle a, b\rangle$ this may be rewritten as:

for all $\varepsilon > 0$ there exists $\delta > 0$ such that $\left|f(\mathbf{x}) - L\right| < \varepsilon$ whenever $0 < \left|\mathbf{x} - \mathbf{a}\right| < \delta$.

This version of the definition may be used for functions of three or more variables without modification. Limit theorems similar to those for functions of one variable hold; for example,

if $\lim\limits_{(x, y)\to(a, b)} f(x, y) = L$ and $\lim\limits_{(x, y)\to(a, b)} g(x, y) = M$,

then $\lim\limits_{(x, y)\to(a, b)} [f(x, y) + g(x, y)] = L + M$.

If $\lim\limits_{(x, y)\to(a, b)} f(x, y) = L$, then $f(x, y)$ approaches L as (x, y) approaches (a, b) along any curve containing (a, b). If two different curves containing (a, b) yield different limits as (x, y) approaches (a, b) along each, then the limit of f does not exist.

1) Evaluate each:

a) $\displaystyle\lim_{(x,y)\to(4,2)} (6x + 3y)$

$6(4) + 3(2) = 30.$

b) $\displaystyle\lim_{(x,y)\to(0,0)} e^{1/(x^2 + y^2)}$

$\infty.$

c) $\displaystyle\lim_{(x,y)\to(3,2)} (y^3 - y)$

$2^3 - 2 = 6.$

d) $\displaystyle\lim_{(x,y)\to(1,-1)} \dfrac{x^2 - 2xy + y^2 - 4}{x - y - 2}$

At $(1, -1)$ we have $\frac{0}{0}$. We can factor and

cancel: $\dfrac{x^2 - 2xy + y^2 - 4}{x - y - 2} = \dfrac{(x - y)^2 - 4}{x - y - 2}$

$\qquad = \dfrac{(x - y + 2)(x - y - 2)}{x - y - 2} = x - y + 2.$

$\displaystyle\lim_{(x,y)\to(1,-1)} x - y + 2 = 1 + 1 + 2 = 4.$

2) Let $f(x, y) = \dfrac{x^3 y}{2x^6 + y^2}$ for $(x, y) \neq (0, 0)$.
Evaluate $\displaystyle\lim_{(x,y)\to(0,0)} f(x, y)$:

a) along the curve $y = 0$.

$f(x, 0) = 0$ for all x so f has limit 0 along $y = 0$.

b) along the curve $y = x^2$.

$f(x, x^2) = \dfrac{x^3 x^2}{2x^6 + x^4} = \dfrac{x}{2x^2 + 1}.$
As $x \to 0$, $\dfrac{x}{2x^2 + 1} \to 0$, so f has limit 0
along $y = x^2$.

c) along the curve $y = x^3$.

$f(x, x^3) = \dfrac{x^3 x^3}{2x^6 + x^6} = \dfrac{1}{3}.$
f has limit $\frac{1}{3}$ along $y = x^3$.

3) Find $\displaystyle\lim_{(x,y)\to(0,0)} f(x, y)$ for the function in
question 2.

By parts b) and c) f approaches different
limits along different curves. Thus the limit
does not exist.

4) True, False:

If $\displaystyle\lim_{(x,y)\to(0,0)} f(x, y) = L$ and
$\displaystyle\lim_{(x,y)\to(0,0)} g(x, y) = M$ then
$\displaystyle\lim_{(x,y)\to(0,0)} [f(x, y)g(x, y)] = LM.$

True.

Page 942 (ET Page 906)

B. The function $z = f(x, y)$ *is continuous at* (a, b) if $\displaystyle\lim_{(x, y)\to(a, b)} f(x, y) = f(a, b)$. f *is continuous on* D if f is continuous at every point (a, b) in D.

Polynomials in x and y are continuous everywhere; rational functions are continuous everywhere they are defined.

5) Where is $f(x, y) = \ln (x - y)$ continuous?

$\ln(x - y)$ is continuous on its domain. The domain is $\{(x, y) \mid x > y\}$.

6) Where is $f(x, y) = \dfrac{2x + y}{x^2 + xy}$ continuous?

Since f is a rational function, f is continuous everywhere on its domain. The domain is all points (x, y) such that $x^2 + xy \neq 0$, i.e., $x(x + y) \neq 0$ so $x \neq 0$ and $x \neq -y$.

7) Is $f(x, y) = \begin{cases} \dfrac{x^3 y}{2x^6 + y^2} & (x, y) \neq (0, 0) \\ 0 & (x, y) = (0, 0) \end{cases}$

continuous at $(0, 0)$?

No. By problem 3) $\displaystyle\lim_{(x, y)\to(0, 0)} f(x, y)$ does not exist.

8) Where is $f(x, y) = e^{\frac{1}{x^2 + y^2}}$ continuous?
Graph f for $0.5 \leq x \leq 2, 0.5 \leq y \leq 2$

$f(x, y)$ is continuous everywhere except at $(0, 0)$.

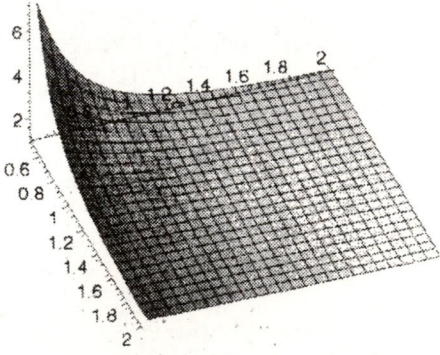

Section 15.3   Partial Derivatives

This section starts the process of finding derivatives for functions of two or more variables. For two variables, we shall see in this section that there is a derivative in the x-direction and another in the y direction and these may be obtained by a process similar to that for functions of one variable. Each of these two "partial" derivatives has, in turn, two partial derivatives; so the original function has four "second derivatives". Clairaut's Theorem says that for some functions two of these four are the same.

Concepts to Master

A. Partial derivatives; Slopes of tangent line; Instantaneous rate of change; Implicit partial differentiation

B. Higher order partial derivatives; Clairaut's Theorem

Summary and Focus Questions

Page 947 (ET Page 911)

A. For $z = f(x, y)$, *the partial derivative of f with respect to x is*
$$f_x(x, y) = \lim_{h \to 0} \frac{f(x + h, y) - f(x, y)}{h}.$$
Since y is unchanging in this definition, $f_x(x, y)$ is computed by treating y as a constant.

For example, if $f(x, y) = x^3y^2$, then $f_x(x, y) = (3x^2)y^2 = 3x^2y^2$.

Other notations for $f_x(x, y)$ include
$$f_x, \ \frac{\partial f}{\partial x}, \ \frac{\partial}{\partial x}f(x, y), \ \frac{\partial z}{\partial x}, \ D_xf, \ D_xf(x, y), \ D_1f, f_1$$

$f_x(a, b)$ may be interpreted as the slope of the tangent line to the surface $z = f(x, y)$ determined by the trace $y = b$. It is the slope at (a, b) in "the x-direction".

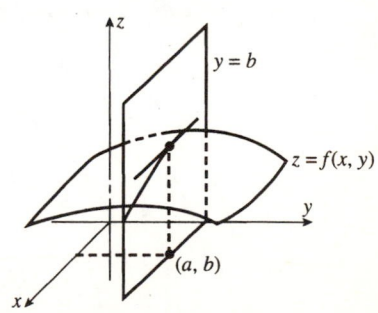

$f_x(x, y)$ is the instantaneous rate of change of f with respect to x.

Likewise, $f_y(x, y) = \lim_{h \to 0} \frac{f(x, y + h) - f(x, y)}{h}$ is computed by holding x constant and is the slope of the tangent line to $z = f(x, y)$ in the y direction.

In a similar manner, partial derivatives may be defined and computed for functions of more than two variables. For example, if
$$f(x, y, z, w) = 2x^2y + 3xz^3w + z^2y^2w,$$

then, treating x, y, and w as constants, we compute

$$\frac{\partial f}{\partial z} = 0 + 3x(3z^2)w + (2z)(y^2w) = 9xz^2w + 2zy^2w.$$

Implicit partial differentiation is performed in the same manner as was done for functions of one variable but remember to treat the other variables as constants. For example, to compute $\frac{\partial f}{\partial x}$ where $z = f(x, y)$ is defined by $4x^2 + 9y^2 + 16z^2 = 100$, we have $8x + 0 + 32z\frac{\partial z}{\partial x} = 0$. Hence $\frac{\partial z}{\partial x} = -\frac{x}{4z}$.

1) For $w = f(x, y, z)$ define $\frac{\partial f}{\partial z}$.

$$\frac{\partial f}{\partial z} = \lim_{h \to 0} \frac{f(x, y, z + h) - f(x, y, z)}{h}.$$

2) Find f_x and f_y for

    a) $f(x, y) = e^{x^2 - y^2}$

$$f_x = e^{x^2 - y^2}(2x) = 2xe^{x^2 - y^2}.$$
$$f_y = e^{x^2 - y^2}(-2y) = -2ye^{x^2 - y^2}.$$

    b) $f(x, y) = xy \sec x$

$$f_x = (xy)(\sec x \tan x) + (\sec x)y$$
$$= y \sec x(x \tan x + 1).$$
$$f_y = x \sec x.$$

3) Find f_y for $f(x, y, z) = \dfrac{x}{\sqrt{y + z}}$.

$$f(x, y, z) = x(y + z)^{-1/2}.$$
$$f_y = x\left(-\frac{1}{2}\right)(y + z)^{-3/2}(1) = -\frac{x}{2(y + z)^{3/2}}.$$

4) Find the slope of the tangent line in the y direction to $f(x, y) = x^2 + 4xy + y^2$ at the point $(1, 2, 13)$.

$$f_y = 4x + 2y.$$
$$f_y(1, 2) = 4(1) + 2(2) = 8.$$

5) The level curves for $f(x, y)$ are given below:

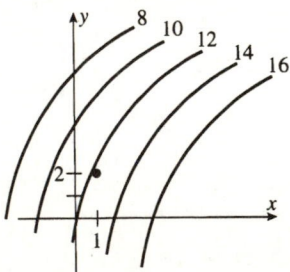

Is $f_x(1, 2)$ positive or negative? What about $f_y(1, 2)$?

As we proceed in the x-direction from $(1, 2)$ the function values get larger. Thus $f_x(1, 2)$ is positive. $f_y(1, 2)$ is negative.

6) Find $\frac{\partial z}{\partial x}$ where $z = f(x, y)$ is given by
$x^2 + y^2 + z^2 = \sin(xyz)$.

Treat y as a constant:

$$2x + 0 + 2z \cdot \frac{\partial z}{\partial x} = \cos(xyz)\left[xy\frac{\partial z}{\partial x} + yz\right].$$

$$2x - yz\cos(xyz) = \frac{\partial z}{\partial x}[xy\cos(xyz) - 2z].$$

$$\frac{\partial z}{\partial x} = \frac{2x - yz\cos(xyz)}{xy\cos(xyz) - 2z}.$$

**Page 951
(ET Page 915)**

B. For a function $z = f(x, y)$ there are four second partial derivatives:

$$f_{xx}(x, y) = \frac{\partial^2 f}{\partial x^2} \qquad f_{yy}(x, y) = \frac{\partial^2 f}{\partial y^2}$$

$$f_{xy}(x, y) = \frac{\partial^2 f}{\partial y \partial x} \qquad f_{yx}(x, y) = \frac{\partial^2 f}{\partial x \partial y}$$

For example, $f_{xy}(x, y)$ is determined by finding the partial derivative of f with respect to x and taking the derivative of that result with respect to y.

Clairaut's Theorem: If f is defined on a disk containing (a, b) and both f_{xy} and f_{yx} are continuous on that disk, then $f_{xy} = f_{yx}$ at (a, b).

7) Find $\frac{\partial^2 f}{\partial x \partial y}$ for $f(x, y) = 3xy^3 + 8x^2y^4$.

$f_y = 9xy^2 + 32x^2y^3$.
Thus $f_{xy} = \frac{\partial^2 f}{\partial x \partial y} = 9y^2 + 64xy^3$.

8) For $f(x, y) = \sin(x^2 + y^2)$, is
$f_{xy}(x, y) = f_{yx}(x, y)$ for all (x, y)?

Yes, because both f_{xy} and $f_{yx}(x, y)$ are continuous.

9) True or False:
If all third partial derivatives are continuous, then

a) $f_{xyz} = f_{zxy}$

True.

b) $f_{xxyz} = f_{xyzy}$

False.

10) Let $f(x, y) = 4x^2y^5 + 3x^3y^2$.

a) Compute f_{xyy}.

$f_x = 8xy^5 + 9x^2y^2$
$f_{xy} = 40xy^4 + 18x^2y$
$f_{xyy} = 160xy^3 + 18x^2$.

b) Compute f_{yxy}.

$f_{yxy} = 160xy^3 + 18x^2$ (same as part a)).

Section 15.4 Tangent Planes and Linear Approximations

This section generalizes the notion of a tangent line for a curve to that of a tangent plane for a surface. The tangent plane is determined by the tangent lines in the x and y directions. Tangents planes will be used to define a differential that may be used to find a linear approximation of a functional value. Differentials and linear approximations generalize to functions with more than two variables.

Concepts to Master

A. Tangent plane to $z = f(x, y)$

B. Differential; Differentiability

Summary and Focus Questions

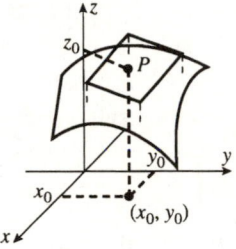

Page 959 (ET Page 923)

A. For a surface given by $z = f(x, y)$, where f has continuous first partial derivatives, all the tangent lines at a given point to a surface form a plane called the *tangent plane*. If $P: (x_0, y_0, z_0)$ is a point on $z = f(x, y)$, the tangent plane at P has equation

$$z - z_0 = f_x(x_0, y_0)(x - x_0) + f_y(x_0, y_0)(y - y_0).$$

This equation has some familiar terms in it. For example, $z - z_0 = f_x(x_0, y_0)(x - x_0)$ is the equation of the tangent line to the curve $z = f(x, y_0)$ (y_0 is constant) in the plane $y = y_0$.

1) Find the equation of the tangent plane to $z = 4x^2y$ at $(1, 3)$.

$f(1, 3) = 4(1)^2 3 = 12.$
$f_x = 8xy, f_x(1, 3) = 8(1)(3) = 24.$
$f_y = 4x^2, f_y(1, 3) = 4(1)^2 = 4.$
The plane is
$z - 12 = 24(x - 1) + 4(y - 3)$, or
$24x + 4y - z = 24.$

2) Where is the tangent plane horizontal for $z = 4x^3 + 3x^2y - 48y$?

Both f_x and f_y must be zero.
$f_x = 12x^2 + 6xy = 6x(2x + y) = 0$
at $x = 0$ or $y = -2x.$
$f_y = 3x^2 - 48 = 0$ at $x^2 = 16.$
Thus $x = 4$ and $x = -4.$
At $x = 4, y = -8$ and $x = -4, y = 8.$

The tangent plane is horizontal at $(4, -8)$ and $(-4, 8).$

Page
962
(ET Page
926)

B. If $z = f(x, y)$ then, similar to functions of one variable, the *change in z* is
defined as $\Delta z = f(x + \Delta x, y + \Delta y) - f(x, y)$.

If we let $dx = \Delta x$ and $dy = \Delta y$ then the *(total) differential* is
$$dz = f_x(x, y)dx + f_y(x, y)dy.$$
The function $z = f(x, y)$ is *differentiable at* (a, b) if
$$\Delta z = f_x(a, b)\Delta x + f_y(a, b)\Delta y + \varepsilon_1\Delta x + \varepsilon_2\Delta y$$
where $\varepsilon_1, \varepsilon_2$ are each functions of Δx and Δy such that $\varepsilon_1 \to 0$ and
$\varepsilon_2 \to 0$ as $(\Delta x, \Delta y) \to (0, 0)$.

The functions ε_1 and ε_2 measure the difference between Δz and dz. Since
their limits are 0 when f is differentiable, dz may be used as a linear
approximation to Δz.

Similar results hold for functions of three or more variables. For example, if
$s = f(w, x, y, z)$, then $ds = f_w dw + f_x dx + f_y dy + f_z dz$.

3) Find Δz and dz for $z = 2x^2 + y^3$ as (x, y)
changes from $(2, 1)$ to $(2.01, 1.03)$.

> At $(2, 1)$, $z = 2(2)^2 + 1^3 = 9$.
>
> At $(2.01, 1.03)$,
> $z = 2(2.01)^2 + (1.03)^3 = 9.1729$.
>
> Thus $\Delta z = 9.1729 - 9 = 0.1729$.
>
> $\frac{\partial z}{\partial x} = 4x$. At $(2, 1)$, $\frac{\partial z}{\partial x} = 4(2) = 8$.
>
> $\frac{\partial z}{\partial y} = 3y^2$. At $(2, 1)$, $\frac{\partial z}{\partial y} = 3(1)^2 = 3$.
>
> $dx = \Delta x = 2.01 - 2 = 0.01$
>
> $dy = \Delta y = 1.03 - 1 = 0.03$
>
> Thus $dz = f_x dx + f_y dy$
> $= 8(0.01) + 3(0.03) = 0.17$.
>
> So dz is within 0.0029 of Δz.

4) Find dz for $z = xe^{xy}$.

> $dz = (xye^{xy} + e^{xy})dx + (x^2 e^{xy})dy$.

5) For $w = f(x, y, z) = xy^3 + yz^3$, find dw.

> $dw = f_x dx + f_y dy + f_z dz$
> $= y^3 dx + (3xy^2 + z^3)dy + 3yz^2 dz$.

6) Estimate $\frac{(3.02)^2}{(0.99)^3}$ using a differential.

Let $z = f(x, y) = x^2 y^{-3}$. We estimate
$f(3.02, 0.99)$ by calculating $f(3, 1)$ and dz.
$f(3, 1) = 3^2(1)^{-3} = 9$.
$f_x = 2xy^{-3} = 6$ at $(3, 1)$.
$f_y = -3x^2 y^{-4} = -27$ at $(3, 1)$. Thus
$dz = f_x dx + f_y dy$
$\quad = 6(0.02) + (-27)(-0.01) = 0.39$.
Finally, $f(3.02, 0.99) = f(3, 1) + \Delta z$
$\quad\quad \approx f(3, 1) + dz = 9 + 0.39 = 9.39$.
$\left(\text{Note: To 4 decimals, } \frac{(3.02)^2}{(0.99)^3} = 9.3995.\right)$

7) True or False:

a) If $f_x(x_0, y_0)$ and $f_y(x_0, y_0)$ exist then f is differentiable at (x_0, y_0).

False. (f_x and f_y must also be continuous.)
A counterexample at $(x_0, y_0) = (0, 0)$ is
$$f(x, y) = \begin{cases} \dfrac{xy}{x^2 + y^2} & (x, y) \neq (0, 0) \\ 0 & (x, y) = (0, 0) \end{cases}$$
Both f_x and f_y are zero at $(0, 0)$ but f is not continuous.

b) If f is differentiable at (x_0, y_0) then $f_x(x_0, y_0)$ and $f_y(x_0, y_0)$ exist.

True.

8) A circular spa is 7 ft. in diameter and 3 ft. deep. If the measurements are accurate to within 0.05 ft. then use differentials to estimate the maximum error in calculating the volume of water.

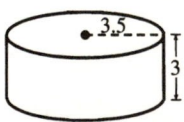

$V = \pi r^2 h$
$dV = \dfrac{\partial V}{\partial r}\, dr + \dfrac{\partial V}{\partial h}\, dh$
$\quad = 2\pi r h\, dr + \pi r^2\, dh$
$\quad \approx 2\pi r h\, \Delta r + \pi r^2\, \Delta h$.
We are given $|\Delta r| \leq 0.05$ and $|\Delta h| \leq 0.05$.
Thus, at $r = 3.5$ and $h = 3$ we have
$dV \approx 2\pi(3.5)(3)(0.05) + \pi(3.5)^2(0.05)$
$\quad = 1.6625\pi \approx 5.22 \text{ ft.}^3$.

Section 15.5   The Chain Rule

The Chain Rule has different versions depending upon the number of variables involved. The Chain Rule also comes in handy for finding derivatives by implicit differentiation.

Concepts to Master

A. Forms of the Chain Rule
B. Implicit Differentiation using the Chain Rule

Summary and Focus Questions

Page 968
(ET Page 932)

A. *The Chain Rule*: Suppose x and y are differentiable functions of t and $z = f(x, y)$ is differentiable. Then z is differentiable with respect to t and
$$\frac{dz}{dt} = \frac{\partial z}{\partial x}\frac{dx}{dt} + \frac{\partial z}{\partial y}\frac{dy}{dt}.$$
Suppose $x = g(s, t)$, $y = h(s, t)$, and $z = f(s, t)$. In this case the version of the Chain Rule for computing the partial derivatives of z with respect to s and t is
$$\frac{\partial z}{\partial s} = \frac{\partial z}{\partial x}\frac{\partial x}{\partial s} + \frac{\partial z}{\partial y}\frac{\partial y}{\partial s} \qquad \frac{\partial z}{\partial t} = \frac{\partial z}{\partial x}\frac{\partial x}{\partial t} + \frac{\partial z}{\partial y}\frac{\partial y}{\partial t}.$$
Both versions of the rule may be generalized. For example, for $w = f(x, y, z)$, $x = h(u, v, s, t)$, $y = k(u, v, s, t)$, $z = m(u, v, s, t)$, the function w has four partial derivatives which may be found by the Chain Rule. For instance,
$$\frac{\partial w}{\partial v} = \frac{\partial w}{\partial x}\frac{\partial x}{\partial v} + \frac{\partial w}{\partial y}\frac{\partial y}{\partial v} + \frac{\partial w}{\partial z}\frac{\partial z}{\partial v}.$$
There are as many terms in the sum as there are intermediate variables (x, y, and z in this case).

1) Find $\frac{dz}{dt}$ where $z = x^2 y$, $x = e^t$, $y = t^2$.

$$\frac{dz}{dt} = \frac{\partial z}{\partial x}\frac{dx}{dt} + \frac{\partial z}{\partial y}\frac{dy}{dt} = (2xy)e^t + x^2(2t)$$
$$= (2e^t t^2)e^t + (e^t)^2 2t = 2t^2 e^{2t} + 2te^{2t}.$$

2) Find $\frac{\partial w}{\partial u}$ at $(u, v) = \left(\frac{\pi}{2}, \frac{\pi}{2}\right)$, where $w = x^2 yz$, $x = uv$, $y = u \sin v$, $z = v \sin u$.

$$\frac{\partial w}{\partial u} = \frac{\partial w}{\partial x}\frac{\partial x}{\partial u} + \frac{\partial w}{\partial y}\frac{\partial y}{\partial u} + \frac{\partial w}{\partial z}\frac{\partial z}{\partial u}$$
$$= (2xyz)v + x^2 z(\sin v) + x^2 y(v \cos u).$$
At $(u, v) = \left(\frac{\pi}{2}, \frac{\pi}{2}\right)$, $x = \frac{\pi^2}{4}$, $y = \frac{\pi}{2}$, $z = \frac{\pi}{2}$.
Therefore
$$\frac{\partial w}{\partial u} = \left(2\frac{\pi^2}{4}\frac{\pi}{2}\frac{\pi}{2}\right)\frac{\pi}{2} + \left(\frac{\pi^2}{4}\right)^2\frac{\pi}{2}(1) + \left(\frac{\pi^2}{4}\right)^2\frac{\pi}{2}(0)$$
$$= \frac{3}{32}\pi^5.$$

Page
972
(ET Page
936)

B. Here is another way to perform implicit differentiation for a function of one variable (Chapter 3): If $y = f(x)$ is defined implicitly by the equation $F(x, y) = 0$, then
$$\frac{dy}{dx} = -\frac{F_x}{F_y}.$$
For example, if y is a function of x defined by $x^2 + y^2 = 1$, then let $F(x, y) = x^2 + y^2 - 1$. The derivative of y is
$$y' = -\frac{F_x}{F_y} = -\frac{2x}{2y} = -\frac{x}{y}.$$
Likewise, if $z = f(x, y)$ is defined implicitly by $F(x, y, z) = 0$, then $\frac{\partial z}{\partial x} = -\frac{F_x}{F_z}$ and $\frac{\partial z}{\partial y} = -\frac{F_y}{F_z}$ and $\frac{\partial z}{\partial y} = -\frac{F_y}{F_z}$.

3) Find y' where y is defined by
$x^2 y^3 + \cos(xy) = 0$.

Let $F(x, y) = x^2 y^3 + \cos(xy)$.
$F_x = 2xy^3 - y \sin(xy)$
$F_y = 3x^2 y^2 - x \sin(xy)$
Thus $y' = -\frac{F_x}{F_y} = \frac{y \sin(xy) - 2xy^3}{3x^2 y^2 - x \sin(xy)}$.

4) Find the equation of the tangent plane to
$x^2 + 2y^2 + z^2 = 6$ at the point $(2, -1, 1)$.

Instead of solving for z and differentiating, find $\frac{\partial z}{\partial x}$ and $\frac{\partial z}{\partial y}$ implicitly.
Let $F(x, y, z) = x^2 + 2y^2 + z^2 - 6$.
$\frac{\partial z}{\partial x} = -\frac{F_x}{F_z} = -\frac{2x}{2z} = -\frac{x}{z}$.
At $(2, -1, 1)$, $\frac{\partial z}{\partial x} = -2$.
$\frac{\partial z}{\partial y} = -\frac{F_y}{F_z} = -\frac{4y}{2z} = -\frac{2y}{z}$.
At $(2, -1, 1)$, $\frac{\partial z}{\partial y} = 2$.
The tangent plane is
$z - 1 = -2(x - 2) + 2(y + 1)$.

Section 15.6 Directional Derivatives and the Gradient Vector

The partial derivatives $f_x(x, y)$ and $f_y(x, y)$ may be thought of as the derivatives in the x direction and y direction, respectively. This section defines the "directional" derivative for an arbitrary direction $\mathbf{u} = \langle a, b \rangle$ in the xy-plane. $f_x(x, y)$ and $f_y(x, y)$ will be special cases where the x direction is specified by $\mathbf{i} = \langle 1, 0 \rangle$ and the y direction by $\mathbf{j} = \langle 0, 1 \rangle$. The vector that points in the direction where the directional derivative is the largest is called the gradient.

Concepts to Master

A. Directional derivative; Gradient; Maximum value of directional derivative

B. Tangent plane to a level surface for $F(x, y, z)$

Summary and Focus Questions

Page 977 (ET Page 941)

A. The *directional derivative* of $f(x, y)$ at (x_0, y_0) in the direction of the unit vector $\mathbf{u} = \langle a, b \rangle$ is

$$D_{\mathbf{u}}f(x_0, y_0) = \lim_{h \to 0} \frac{f(x_0 + ah, y_0 + bh) - f(x_0, y_0)}{h} = f_x(x_0, y_0)a + f_y(x_0, y_0)b.$$

$D_{\mathbf{u}}f(x_0, y_0)$ may be interpreted as the slope of the tangent line to the graph of $z = f(x, y)$ in the $\mathbf{u}$ direction. It is also the instantaneous rate of change of z at (x_0, y_0) in the $\mathbf{u}$ direction.

The *gradient of $f(x, y)$* at (x_0, y_0) is $\nabla f(x_0, y_0) = \langle f_x(x_0, y_0), f_y(x_0, y_0) \rangle$.

Using the gradient, a directional derivative may be written

$$D_{\mathbf{u}}f(x_0, y_0) = \nabla f(x_0, y_0) \cdot \mathbf{u}. \quad \text{(dot product of } \nabla f \text{ and } \mathbf{u}\text{)}$$

Among all the values for $D_{\mathbf{u}}f(x_0, y_0)$, the maximum occurs in the direction of the gradient

$$\mathbf{u} = \frac{\nabla f(x_0, y_0)}{|\nabla f(x_0, y_0)|}$$

and the maximum value is $|\nabla f(x_0, y_0)|$; that is, $z = f(x, y)$ increases most rapidly in the direction of the gradient. Likewise, $-\nabla f(x_0, y_0)$ is the direction in the xy-plane in which f decreases most rapidly.

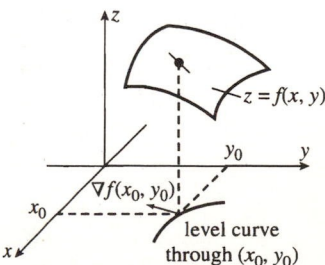

Directional derivatives and gradients for functions of three or more variables are defined similarly.

1) If $\mathbf{u} = \mathbf{i}$, then what is $D_{\mathbf{u}}f(x_0, y_0)$?

The partial derivative $f_x(x_0, y_0)$.

2) Find the directional derivative of

a) $f(x, y) = x^2y^2 - xy^3$ in the direction $\mathbf{u} = \frac{1}{2}\mathbf{i} - \frac{\sqrt{3}}{2}\mathbf{j}$.

$f_x = 2xy^2 - y^3$,
$f_y = 2x^2y - 3xy^2$.
$D_{\mathbf{u}}f = \frac{1}{2}(2xy^2 - y^3) - \frac{\sqrt{3}}{2}(2x^2y - 3xy^2)$
$= \frac{2 + 3\sqrt{3}}{2}xy^2 - \frac{1}{2}y^3 - \sqrt{3}x^2y.$

b) $f(x, y, z) = xy \sin z$ in the direction $\mathbf{u} = \left\langle \frac{1}{2}, -\frac{1}{\sqrt{2}}, \frac{1}{2} \right\rangle$ at the point $\left(2, 1, \frac{\pi}{6}\right)$.

$f_x = y \sin z, f_y = x \sin z$, and $f_z = xy \cos z$.
At $\left(2, 1, \frac{\pi}{6}\right), f_x\left(2, 1, \frac{\pi}{6}\right) = 1\left(\frac{1}{2}\right) = \frac{1}{2}$,
$f_y\left(2, 1, \frac{\pi}{6}\right) = 2\left(\frac{1}{2}\right) = 1$, and
$f_z\left(2, 1, \frac{\pi}{6}\right) = 2(1)\frac{\sqrt{3}}{2} = \sqrt{3}.$
$\nabla f\left(2, 1, \frac{\pi}{6}\right) = \left\langle \frac{1}{2}, 1, \sqrt{3} \right\rangle$ and
$D_{\mathbf{u}}f\left(2, 1, \frac{\pi}{6}\right) = \left\langle \frac{1}{2}, 1, \sqrt{3} \right\rangle \cdot \left\langle \frac{1}{2}, -\frac{1}{\sqrt{2}}, \frac{1}{2} \right\rangle$
$= \frac{1}{2}\left(\frac{1}{2}\right) + 1\left(-\frac{1}{\sqrt{2}}\right) + \sqrt{3}\left(\frac{1}{2}\right)$
$= \frac{1 - 2\sqrt{2} + 2\sqrt{3}}{4}.$

3) Find the gradient of $f(x, y) = e^{xy}$.

$f_x = ye^{xy}$ and $f_y = xe^{xy}$, so
$\nabla f(x, y) = ye^{xy}\mathbf{i} + xe^{xy}\mathbf{j}.$

4) Let $f(x, y, z) = xy^2 + yz^2$.

a) What is the maximum value of $D_{\mathbf{u}}f(1, 2, 1)$ as $\mathbf{u}$ varies?

The maximum value is $|\nabla f(1, 2, 1)|$.
$\nabla f = \langle y^2, 2xy + z^2, 2yz \rangle.$
Thus $\nabla f(1, 2, 1) = \langle 4, 5, 4 \rangle$ and
$|\nabla f(1, 2, 1)| = \sqrt{4^2 + 5^2 + 4^2} = \sqrt{57}.$

b) What direction $\mathbf{u}$ gives the maximum value of $D_{\mathbf{u}}f(1, 2, 1)$?

$\mathbf{u}$ is the direction $\nabla f(1, 2, 1)$. But $\mathbf{u}$ must be a unit vector, so $\mathbf{u} = \left\langle \frac{4}{\sqrt{57}}, \frac{5}{\sqrt{57}}, \frac{4}{\sqrt{57}} \right\rangle.$

c) Find $D_{\mathbf{v}}f(1, 2, 1)$ where $\mathbf{v} = \langle \frac{5}{6}, \frac{1}{2}, \frac{\sqrt{2}}{6} \rangle$.

$$D_{\mathbf{v}}f(1, 2, 1) = \nabla f(1, 2, 1) \cdot \mathbf{v}$$
$$= 4\left(\frac{5}{6}\right) + 5\left(\frac{1}{2}\right) + 4\left(\frac{\sqrt{2}}{6}\right) = \frac{35 + 4\sqrt{2}}{6}.$$

Page 983 (ET Page 947)

B. Let S be a level surface $F(x, y, z) = k$ and $P(x_0, y_0, z_0)$ be a point of S. The gradient $\nabla F(x_0, y_0, z_0)$ is normal to the surface S at point P. Thus the tangent plane to S at P has equation
$$\nabla P(x_0, y_0, z_0) \cdot \langle x - x_0, y - y_0, z - z_0 \rangle = 0,$$
or
$$F_x(x_0, y_0, z_0)(x - x_0) + F_y(x_0, y_0, z_0)(y - y_0)$$
$$+ F_z(x_0, y_0, z_0)(z - z_0) = 0.$$

Likewise, the line normal to the surface at P_0 has equation
$$\frac{x - x_0}{F_x(x_0, y_0, z_0)} = \frac{y - y_0}{F_y(x_0, y_0, z_0)} = \frac{z - z_0}{F_z(x_0, y_0, z_0)}.$$
In the two dimensional case, $z = f(x, y)$, the gradient of $f(x, y)$ at (x_0, y_0) is a vector in the xy-plane perpendicular to the level curve of f at (x_0, y_0). This agrees with the discussion in part A where we saw that $\nabla f(x_0, y_0)$ is the direction in which the increase in $f(x, y)$ is greatest.

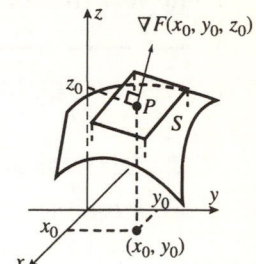

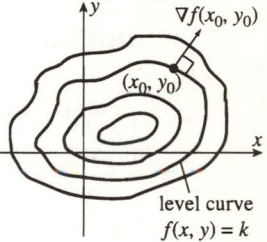

level curve
$f(x, y) = k$

5) Find the equations of the tangent plane and normal line to $x^2 + 2y^2 - z^2 - 4xyz = 16$ at $(1, 4, 1)$.

Let $F(x, y, z) = x^2 + 2y^2 - z^2 - 4xyz - 16$.
$\nabla F = \langle 2x - 4yz, 4y - 4xz, -2z - 4xy \rangle$.
$\nabla F(1, 4, 1) = \langle -14, 12, -18 \rangle$.

The tangent plane is
$-14(x - 1) + 12(y - 4) - 18(z - 1) = 0$,
$7x - 6y + 9z = -8$.

The normal line is $\frac{x - 1}{-14} = \frac{y - 4}{12} = \frac{z - 1}{-18}$,
or $\frac{x - 1}{7} = \frac{4 - y}{6} = \frac{z - 1}{9}$.

6) Given the level curve to $z = f(x, y)$ below, which unit vectors below could be gradients?

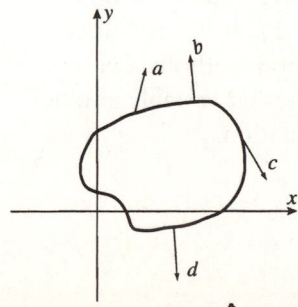

b and **d** appear to be normal to the level curve and are either gradients or the negatives of gradients.

Section 15.7   Maximum and Minimum Values

In this section you will learn how to use partial derivatives to find the maxima and minima of functions of two variables. The terms, concepts, and procedures are very similar to those for functions of one variable.

Concepts to Master

A. Local maximum; Local minimum; Critical point; Second Derivative Test

B. Absolute maximum and minimum; Absolute extrema on a closed, bounded set; Extreme value problems

Summary and Focus Questions

Page 989
(ET Page 953)

A. A function $z = f(x, y)$ has a *local maximum at* (a, b) if $f(a, b) \geq f(x, y)$ for all (x, y) in some open disk about (a, b). A *local minimum at* (a, b) is defined similarly but with $f(a, b) \leq f(x, y)$. These are *local extrema* of f.

A *critical point* (a, b) of f is a point in the domain of f for which both $f_x(a, b) = 0$ and $f_y(a, b) = 0$ or at least one of the partial derivatives does not exist.

If (a, b) is a local extremum for a continuous $z = f(x, y)$ then (a, b) is a critical point. The converse is false. A critical point that is not a local extremum is a *saddle point*.

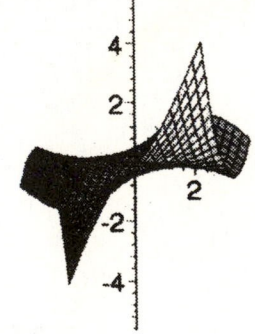

For $f(x, y) = x^2y + xy^2$ both $f_x(x, y) = 2xy + y^2$ and $f_y(x, y) = x^2 + 2xy$ are zero at $(0, 0)$. Thus $(0, 0)$ is a critical point. However, along the line $y = 2x, f(x, y)$ becomes $f(x, y) = x^2(2x) + x(2x)^2 = 6x^3$. Thus $(0, 0)$ is not a local maximum or local minimum. the graph of f is at the right.

The method of finding some critical points for $z = f(x, y)$ involves setting $f_x = 0$ and $f_y = 0$ and solving these two equations in two variables simultaneously. Often the system $f_x = 0$ and $f_y = 0$ is not linear and may be very difficult to solve. There are no general methods. One approach is to solve for one variable in terms of the other variable in one equation and substitute that value into the other equation.

Second Derivative Test: Suppose $z = f(x, y)$ has continuous second partial derivatives in a disk with center (a, b). Suppose $f_x(a, b) = 0$ and $f_y(a, b) = 0$ and let $D(a, b) = f_{xx}(a, b)f_{yy}(a, b) - [f_{xy}(a, b)]^2$.

Then the following holds:

Conditions	Conclusion
$D > 0, f_{xx} > 0$	f has a local minimum at (a, b)
$D > 0, f_{xx} < 0$	f has a local maximum at (a, b)
$D < 0$	f has a saddle point at (a, b)
$D = 0$	No conclusion can be made

A shorthand way to remember D is as a determinant:

$$D = \begin{vmatrix} f_{xx} & f_{xy} \\ f_{yx} & f_{yy} \end{vmatrix} = f_{xx}f_{yy} - (f_{xy})^2$$

1) Let $f(x, y) = y^3 - 24x - 3x^2y$.

   a) Find the critical points of f.

Since f is a polynomial, critical points occur only where $f_x = 0$ and $f_y = 0$:
$$f_x = -24 - 6xy = 0$$
$$f_y = 3y^2 - 3x^2 = 0$$

To solve this, we observe from the second equation that $3x^2 = 3y^2$, so $x = y$ or $x = -y$. Substituting $x = y$ in the first:
$$-24 - 6x(x) = 0$$
$$-24 - 6x^2 = 0$$
$$6x^2 = -24$$

This has no solution.

Substituting $x = -y$ in the first:
$$-24 - 6x(-x) = 0$$
$$-24 + 6x^2 = 0$$
$$6x^2 = 24$$
$$x^2 = 4; x = 2, -2$$

When $x = 2, y = -2$ and when $x = -2$, $y = 2$. There are no points where f_x or f_y does not exist. Thus the critical points are $(2, -2)$ and $(-2, 2)$.

   b) Find the local extrema of $f(x, y)$.

The extrema are among the critical points.
$$f_{xx} = -6y, f_{xy} = -6x, f_{yy} = 6y.$$
$$D = (-6y)(6y) - (-6x)^2 = -36y^2 - 36x^2$$
At both $(2, -2)$ and $(-2, 2)$,
$$D = -288 < 0.$$

Thus both $(2, -2)$ and $(-2, 2)$ are saddle points. There are no local extrema.

2) Suppose f has continuous second derivatives and has four critical points with the following information about the second derivatives.

Point	f_{xx}	f_{yy}	f_{xy}
A: (7, 1)	8	2	−4
B: (1, 2)	1	9	2
C: (0, 4)	−6	3	1
E: (−1, 3)	−2	−5	3

Classify each critical point.

We calculate $D = f_{xx}f_{yy} - (f_{xy})^2$ for each:

Point	D
A	$8(2) - (-4)^2 = 0$
B	$1(9) - 2^2 = 5$
C	$-6(3) - 1^2 = -19$
E	$-2(-5) - 3^2 = 1$

By the Second Derivative Test:
B is a local minimum, C is a saddle point, E is a local maximum.

We cannot conclude anything about A from the information given.

3) **a)**   Find all critical points of
$$f(x,y) = x^2 - 4xy + 2x - \frac{4}{15}(5y + 41)^{3/2}.$$

$f_x = 2x - 4y + 2 = 2(x - 2y + 1).$

$f_y = -4x - \frac{4}{15}\left(\frac{3}{2}\right)(5y + 41)^{1/2}(5)$
$\quad = -2(2x + \sqrt{5y + 41}).$

Set both f_x and f_y to 0.

$x - 2y + 1 = 0 \qquad 2x + \sqrt{5y + 41} = 0.$

Solve for x in the first:

$x = 2y - 1$

and substitute:

$2(2y - 1) + \sqrt{5y + 41} = 0$
$2 - 4y = \sqrt{5y + 41}$
$4 - 16y + 16y^2 = 5y + 41$
$16y^2 - 21y - 37 = 0$
$(16y - 37)(y + 1) = 0$
$y = \frac{37}{16}, y = -1.$

From $x = 2y - 1$, the possible critical points are $\left(\frac{29}{8}, \frac{37}{16}\right)$ and $(-3, -1)$.

$f_x(-3, -1) = f_y(-3, -1) = 0$ but $f_y\left(\frac{29}{8}, \frac{37}{16}\right) \neq 0$. Thus $(-3, -1)$ is the only critical point.

b) Classify the critical point in part a).

$f_{xx} = 2, f_{xy} = -4$ and
$f_{yy} = \frac{1}{2}(5y + 41)^{-1/2}(5) = \dfrac{5}{2\sqrt{5y + 41}}$.
At $(-3, -1), f_{yy} = \frac{5}{12}$.

$f_{xx}f_{yy} - (f_{xy})^2 = 2\left(\frac{5}{12}\right) - (-4)^2$, which is negative. Thus $(-3, -1)$ is a saddle point.

B. The *absolute maximum* of f is the value $f(a, b)$ for some (a, b) such that $f(a, b) \geq f(x, y)$ for all (x, y) in the domain of f. *Absolute minimum* is defined as above but with $f(a, b) \leq f(x, y)$. These are the *extreme values* of f.

Page 995 (ET Page 959)

A continuous function whose domain is closed (contains all its boundary points) and bounded has an absolute maximum and absolute minimum.

To find the absolute extrema of continuous $f(x, y)$ on a closed and bounded set D, compute $f(x, y)$ at critical points in D and along the boundary of D.

4) Does the absolute extrema exist for $f(x, y) = x^2 + y^2$ with domain $D = \{(x, y)|\ 0 \leq x \leq 2, 0 \leq y \leq 3\}$?

Yes, f is continuous and D is both closed and bounded. (D is a rectangle.)

5) Find the extreme values of $f(x, y) = 10 - x^2 - 2y^2 + 2x + 8y$ with domain D as graphed.

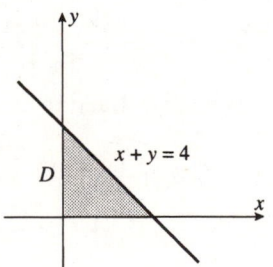

First find the critical points:
$f_x = -2x + 2 = 0$, at $x = 1$.
$f_y = -4y + 8 = 0$, at $y = 2$.

$(1, 2)$ is the only critical point and it is within D. D has three boundary lines:
l_1(x-axis): $y = 0, 0 \leq x \leq 4$.

Here $f(x, y) = f(x, 0) = 10 - x^2 + 2x$ has a maximum at $(1, 0)$ and a minimum at $(4, 0)$.

l_2(y-axis): $x = 0, 0 \leq y \leq 4$.

Here $f(x, y) = f(0, y) = 10 - 2y^2 + 8y$ has a maximum at $(0, 2)$ and a minimum at $(0, 0)$ and $(0, 4)$.

$l_3(x + y = 4)$: $y = 4 - x, 0 \leq x \leq 4$.

Here $f(x, y) = f(x, 4 - x)$
$= 10 - x^2 - 2(4 - x)^2 + 2x + 8(4 - x)$
$= 10 + 10x - 3x^2$

has a maximum at $x = \frac{5}{3}$, $y = \frac{7}{3}$, and a minimum at $(4, 0)$.

We compute the corresponding $f(x, y)$ values and summarize in a table:

(x, y)	How found?	f(x, y)
(1, 0)	max on l_1	11
(4, 0)	min on both l_1 and l_3	2
(0, 2)	max on l_2	18
(0, 4)	min on l_2	10
$\left(\frac{5}{3}, \frac{7}{3}\right)$	max on l_3	18.33
(1, 2)	critical point	19

The absolute maximum is 19 and occurs at $(1, 2)$. The absolute minimum is 2 and occurs at $(4, 0)$.

6) Show that among all rectangular parallelepipeds with volume 1 cubic inch, the one with smallest surface area is a cube.

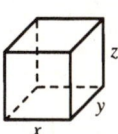

We are given $xyz = 1$ and must minimize $S = 2xy + 2xz + 2yz$.

From $xyz = 1$, $z = \frac{1}{xy}$.

Thus $S = 2xy + 2x\left(\frac{1}{xy}\right) + 2y\left(\frac{1}{xy}\right)$

$= 2xy + \frac{2}{y} + \frac{2}{x}$.

$S_x = 2y - \frac{2}{x^2} = 0$

$2y = \frac{2}{x^2}, y = \frac{1}{x^2}$.

$S_y = 2x - \frac{2}{y^2} = 0$

$2x = \frac{2}{y^2}, x = \frac{1}{y^2}$.

From $y = \frac{1}{x^2}$ and $x = \frac{1}{y^2}$, $y = y^4$. Thus $y = 0$ or $y = 1$. $y = 0$ is not possible so $y = 1$.

Thus $x = \frac{1}{y^2} = 1$ and $z = \frac{1}{xy} = \frac{1}{1 \cdot 1} = 1$.

Therefore the object is a cube with edges 1 inch long.

Section 15.8   **Lagrange Multipliers**

Maximum and minimum value problems are frequently stated in terms of finding the maximum (or minimum) of a function given a constraint on the variables in the form of an equation. For example, in problem 6 of section 15.7 we found the minimum of $2xy + 2xy + 2yz$ under the condition that $xyz = 1$. The solution there involved solving for one variable in terms of the others, substituting that into the function and setting partials equal to zero. This section gives a different method for these same type of problems.

Concepts to Master

Solution to extreme value problems using Lagrange Multipliers

Summary and Focus Questions

Page 1002 (ET Page 966)

If $f(x, y)$ and $g(x, y)$ have continuous partial derivatives and (a, b) is a local extremum for f when restricted to $g(x, y) = k$ (a constraint), then there is a number λ called a *Lagrange multiplier* such that

$$\nabla f(a, b) = \lambda \nabla g(a, b).$$

Thus solving the constrained extremum problem above is the same as solving the equations $\nabla f = \lambda \nabla g$ and $g(x, y) = k$.

The solution to the constrained optimum problem involving three variables is determined by solving $\nabla f(x, y, z) = \lambda \nabla g(x, y, z)$ and $g(x, y, z) = k$.

Example: Find the minimum of $f(x, y, z) = 2xy + 2xz + 2yz$ subject to $xyz = 1$.

$$\nabla f(x, y, z) = \langle 2y + 2z, 2x + 2z, 2x + 2y \rangle$$
$$\nabla g(x, y, z) = \langle yz, xz, xy \rangle$$

From $\nabla f(x, y, z) = \lambda \nabla g(x, y, z)$

$$\begin{array}{lll} 2y + 2z = \lambda yz & & 2xy + 2xz = \lambda xyz \\ 2x + 2z = \lambda xz & \rightarrow & 2xy + 2yz = \lambda xyz \\ 2x + 2y = \lambda xy & & 2xz + 2yz = \lambda xyz \end{array}$$

From the first and second equations $2xz = 2yz$, and, therefore, $x = y$. From the second and third equations, $2xy = 2xz$, and, therefore $y = z$.
Thus $x = y = z$. Since $xyz = 1$, we have $x = 1, y = 1$, and $z = 1$.

For problems with two constraints $g_1(x, y, z) = k_1, g_2(x, y, z) = k_2$:
$$\nabla f(x, y, z) = \lambda_1 \nabla g_1(x, y, z) + \lambda_2 \nabla g_2(x, y, z).$$

1) Find the extrema of
$f(x, y) = x^2 - 4xy + 2y^2$ subject to
$2x - 3y = 14$.

$\nabla f = \langle 2x - 4y, -4x + 4y \rangle$.

Let $g(x, y) = 2x - 3y$.

Then $\nabla g = \langle 2, -3 \rangle$.

From $\nabla f = \lambda \nabla g, g(x, y) = 14$, we have three equations with variables x, y, λ.

(1) $2x - 4y = 2\lambda$

(2) $-4x + 4y = -3\lambda$

(3) $2x - 3y = 14$.

Add (1) and (2) to get $-2x = -\lambda, x = \frac{1}{2}\lambda$

Multiply (1) by 2 and add (2):

$-4y = \lambda, y = -\frac{1}{4}\lambda$.

From (3), $2\left(\frac{1}{2}\lambda\right) - 3\left(-\frac{1}{4}\lambda\right) = 14$,

$\frac{7}{4}\lambda = 14, \lambda = 8$.

Thus $x = \frac{1}{2}\lambda = 4, y = -\frac{1}{4}\lambda = -2$.

$f(4, -2) = 56$.

If $2x - 3y = 14$, then $y = \frac{2}{3}x - \frac{14}{3}$. Thus for large values of x, y is also large. For large values of x (and therefore y) $f(x, y)$ is negative. For large, negative x, $f(x, y)$ is negative. There is no absolute minimum; 56 is the absolute maximum.

2) Find the system of equations for solving:
Maximize
$f(x, y, z) = x^2 + 6xy^2 + yz^2 + yz^3$ subject to
$x + y + z^2 = 2$ and $x^2 + y + z = 3$.

$\nabla f =$
$\langle 2x + 6y^2 + 3z^2, 12xy + z^3, 6xz + 3yz^2 \rangle$.

Let $g_1(x, y, z) = x + y + z^2, \nabla g_1 = \langle 1, 1, 2z \rangle$
and $g_2(x, y, z) = x^2 + y + z, \nabla g_2 = \langle 2x, 1, 1 \rangle$.

The system of five equations with variables x, y, z, λ_1 and λ_2 from $\nabla f = \lambda_1 \nabla g_1 + \lambda_2 \nabla g_2$, $g_1(x, y, z) = 2, g_2(x, y, z) = 3$ is

$2x + 6y^2 + 3z^2 = \lambda_1 + 2x\lambda_2$

$12xy + z^3 = \lambda_1 + \lambda_2$

$6xz + 3yz^2 = 2z\lambda_1 + \lambda_2$

$x + y + z^2 = 2$

$x^2 + y + z = 3$.

3) Find the volume of the largest rectangular box in the first octant with three faces in the coordinate planes and one vertex in the plane $3x + y + 2z = 12$.

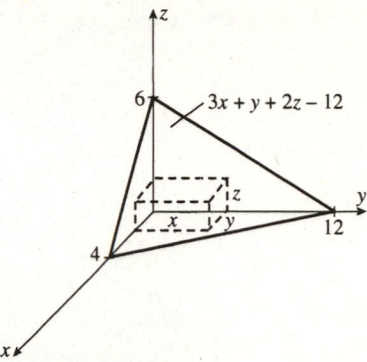

We need to find the maximum volume of the box, $V = xyz$, subject to $3x + y + 2z = 12$.

Let $g(x, y, z) = 3x + y + 2z$.
$\nabla V(x, y, z) = \langle yz, xz, xy \rangle$
$\nabla g(x, y, z) = \langle 3, 1, 2 \rangle$
Thus $\nabla f = \lambda \nabla g$ becomes

$$yz = 3\lambda$$
$$xz = \lambda$$
$$xy = 2\lambda$$

Multiply by x, y, z respectively:

$$xyz = 3x\lambda$$
$$xyz = y\lambda$$
$$xyz = 2z\lambda$$

Therefore $3x\lambda = y\lambda = 2z\lambda$ and for $\lambda \neq 0$

$$3x = y = 2z.$$

Then $3x + y + 2z = 12$ becomes
$y + y + y = 12$, so $y = 4$.

Thus $3x = 4$ or $x = \frac{4}{3}$ and $2z = 4$ or $z = 2$.
The maximum volume occurs when the box is $\frac{4}{3} \times 4 \times 2$.

Technology Plus for Chapter 15

1) Let $f(x, y) = 2xy$.
Use a computer to

a) draw level curves for $c = 0.25, 0.5,$ and 1.

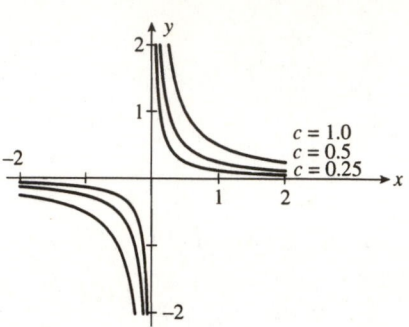

b) draw level curves for $c = -0.25, -0.5,$ and -1.

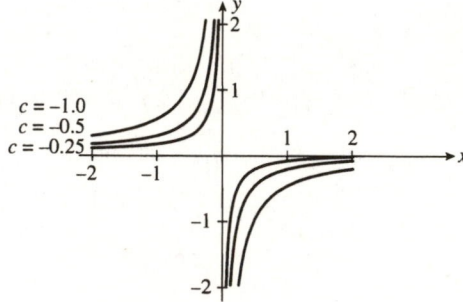

c) graph $f(x, y)$ for $-1 \le x \le 1$, $-1 \le y \le 1$.

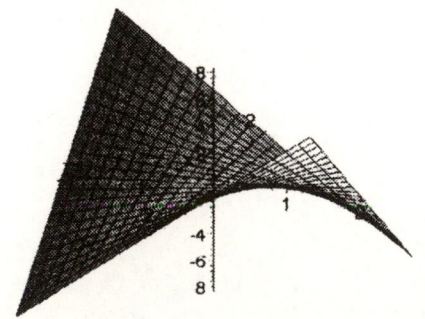

2) Let $z = \sqrt{9 - x^2 - y^2}$.

    a) Find $f_x(1, 2)$ and the parametric equations for the tangent line to f at $(1, 2)$ in the x-direction.

$$f_x(x, y) = \tfrac{1}{2}(9 - x^2 - y^2)^{-1/2}(-2x)$$
$$= \frac{-x}{\sqrt{9 - x^2 - y^2}}.$$
$$f_x(1, 2) = -\tfrac{1}{2}.$$

$f(1, 2) = 2$. Therefore, the line is
$$x = 1 + y$$
$$y = 2$$
$$z = 2 - \tfrac{1}{2}t.$$

    b) Find $f_y(1, 2)$ and the parametric equations for the tangent line to f at $(1, 2)$ in the y-direction.

$$f_y(x, y) = \tfrac{1}{2}(9 - x^2 - y^2)^{-1/2}(-2y)$$
$$= \frac{-y}{\sqrt{9 - x^2 - y^2}}.$$
$$f_y(1, 2) = \frac{-2}{2} = -1.$$

$$x = 1$$
$$y = 2 + t$$
$$z = 2 - t.$$

    c) On the same screen draw the graphs of $f(x, y)$ and the lines in parts a) and b).

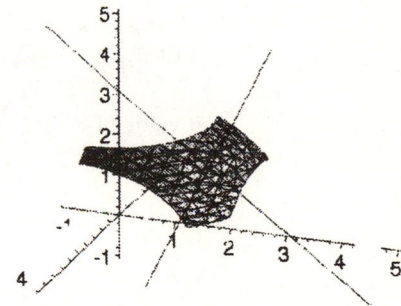

3) a) Use Lagrange multipliers to solve "maximize xy subject to $\frac{x^2}{4} + y^2 = 1$."

Use $x \geq 0, y \geq 0$.

For $f(x, y) = xy$, $\nabla f = \langle y, x \rangle$.

For $g(x, y) = \frac{x^2}{4} + y^2$, $\nabla g = \left\langle \frac{x}{2}, 2y \right\rangle$.

From $\nabla f = \lambda \nabla g$,

$$y = \lambda\left(\frac{x}{2}\right)$$
$$x = \lambda(2y).$$

Multiply the first equation by x and the second by y:

$$xy = \lambda\frac{x^2}{2}$$
$$xy = \lambda 2y^2$$

Thus $\lambda\frac{x^2}{2} = \lambda 2y^2$.

Hence $\frac{x^2}{2} = 2y^2$, $x^2 = 4y^2$, and $x = 2y$. (Note $x = -2y$ would result in a minimum.)

$$\frac{(2y)^2}{4} + y^2 = 1$$
$$2y^2 = 1, \text{ so } y = \frac{1}{\sqrt{2}}.$$

Therefore $x = 2\left(\frac{1}{\sqrt{2}}\right) = \sqrt{2}$ and the maximum of xy is $(\sqrt{2})\left(\frac{1}{\sqrt{2}}\right) = 1$.

b) Confirm your results in part a) by graphing on the same screen

$$\frac{x^2}{4} + y^2 = 1$$

$xy = k$ for $k = 0.33, 0.67, 1, 1.33,$ and $1.67.$

For what value of k does $xy = k$ just touch the ellipse?

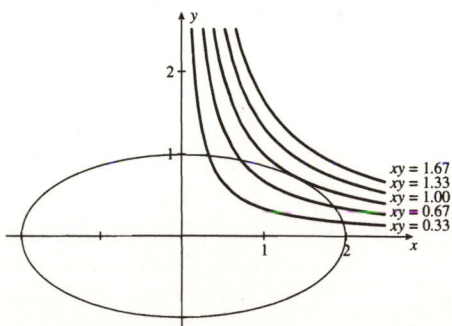

The curve $xy = 1$ just touches the ellipse. Note that 1 is the maximum value of xy subject to $\frac{x^2}{4} + y^2 = 1$.

Chapter 16 — Multiple Integrals

"SORRY I'M LATE — I WAS WORKING OUT PI TO 5,000 PLACES."

Cartoons courtesy of Sidney Harris. Used by permission.

Section 16.1   Double Integrals of Rectangles

Recall that definite integrals are the limits of Riemann sums and may be used to determine the area under a curve. This section defines double integrals as the limits of "double" Riemann sums and uses them to find the volume of a solid under a surface.

Concepts to Master

A. Double Riemann sum; Double integral; Integrable

B. Midpoint Rule; Average value of a function over a rectangle

C. Interpretation of the double integral as a volume; Double integral properties

---

Summary and Focus Questions

Page 1017 (ET Page 981)*

A. For a non-negative function $f(x, y)$ defined on a rectangular region

$$R = [a, b] \times [c, d] = \{(x, y): a \le x \le b, c \le y \le d\}$$

we divide the interval $[a, b]$ into m subintervals of width $\Delta x = \dfrac{b - a}{m}$ and the interval $[c, d]$ into n subintervals of width $\Delta y = \dfrac{d - c}{n}$. The region R is thus divided into subrectangles, R_{ij}, each of which has area $\Delta A = \Delta x \Delta y$. For each R_{ij}, choose a sample point $(x_{ij}{}^*, y_{ij}{}^*)$ from the subrectangle R_{ij}. The resulting *double Riemann sum* is

$$\sum_{i=1}^{m} \sum_{j=1}^{n} f(x_{ij}{}^*, y_{ij}{}^*) \Delta A.$$

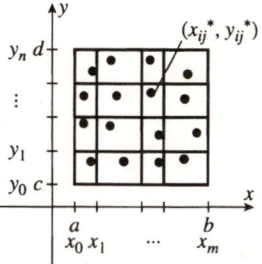

The *double integral of f over R* is the limit of Riemann sums:

$$\iint\limits_{R} f(x, y)\,dA = \lim_{m, n \to \infty} \sum_{i=1}^{m} \sum_{j=1}^{n} f(x_{ij}{}^*, y_{ij}{}^*) \Delta A.$$

If this limit exists, then f is *integrable* over R.

---

1) Find the Riemann sum for $f(x, y) = x^2 + y^2$ over $R = [1, 9] \times [-1, 3]$ with $m = 4$ and $n = 2$. Use the left side midpoint of each subrectangle.

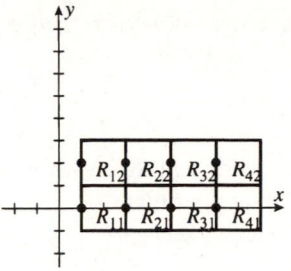

	x_{ij}^*	y_{ij}^*	$f(x_{ij}^*, y_{ij}^*)$
R_{11}	1	0	1
R_{12}	3	0	9
R_{13}	5	0	25
R_{14}	7	0	49
R_{21}	1	2	5
R_{22}	3	2	13
R_{23}	5	2	29
R_{24}	7	2	53
			184

$\Delta A = \Delta x \Delta y = 2(2) = 4$.
The Riemann sum is $184(4) = 736$.

2) For a given set of subrectangles, what choice for (x_{ij}^*, y_{ij}^*) yields the largest Riemann sum for $\iint_R e^{x-y} dA$?

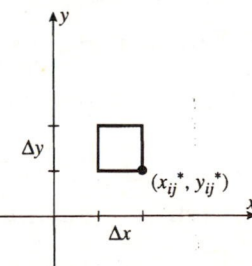

e^{x-y} is largest for large x and small y.
Choose the lower right corner for (x_{ij}^*, y_{ij}^*).

Page
1021
(ET Page
985)

B. *Midpoint Rule:*

If $\bar{x}_i$ is the midpoint of $[x_{i-1}, x_i]$ and $\bar{y}_j$ is the midpoint of $[y_{j-1}, y_j]$, then

$$\iint\limits_R f(x, y)dA \approx \sum_{i=1}^{m} \sum_{j=1}^{n} f(\bar{x}_i^*, \bar{y}_j^*)\Delta A.$$

The *average value* of $f(x, y)$ over a rectangular region R is

$$f_{ave} = \frac{1}{(b-a)(d-c)} \iint\limits_R f(x, y)dA.$$

3) Use the Midpoint Rule with $m = 2$ and $n = 3$ to approximate

$\iint\limits_R xy\, dA$, where R is $[1, 2] \times [2, 5]$.

$\Delta A = \Delta x \Delta y = (0.5)(1) = 0.5.$

	$f(x, y)$		$\bar{y}_i$	
		2.5	3.5	4.5
$\bar{x}_i$	1.25	3.125	4.375	5.625
	1.75	4.375	6.125	7.875

$3.125 + 4.375 + 5.625 + 4.375 + 6.125$
$+ 7.875 = 31.50.$

The Riemann sum is $(0.5)(31.5) = 15.75.$

4) If $R = [1, 2] \times [3, 5]$ and

$\iint\limits_R (x + y)dA = 11$, what is the average

value of $f(x + y) = x + y$ over R?

$$f_{ave} = \frac{1}{(2-1)(5-3)}(11) = \frac{11}{2}.$$

Page
1019
(ET Page
983)

C. Let $R = [a, b] \times [c, d]$. For $f(x, y) \geq 0$ the double

integral $\iint\limits_R f(x, y)dA$ may be interpreted as the volume

of the solid above R and under the surface $z = f(x, y).$

Some properties of double integrals include:

$$\iint\limits_R [f(x, y) + g(x, y)]dA = \iint\limits_R f(x, y)dA + \iint\limits_R g(x, y)dA.$$

$$\iint\limits_R cf(x, y)dA = c\iint\limits_R f(x, y)dA.$$

If $f(x, y) \geq g(x, y)$ for all $(x, y) \in R$, then $\iint\limits_R f(x, y)dA \geq \iint\limits_R g(x, y)dA.$

5) Write a double integral expression for the volume of the pictured solid.

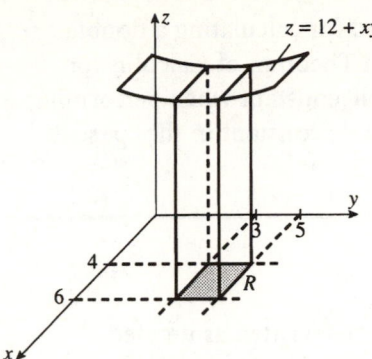

Let R be the rectangular region $4 \le x \le 6$, $3 \le y \le 5$. The volume pictured is

$$\iint_R (12 + xy)\, dA.$$

6) Let $R = [1, 4] \times [5, 9]$. Evaluate $\displaystyle\iint_R 2\, dA.$

$\displaystyle\iint_R 2\, dA$ is the volume of the solid under $f(x, y) = 2$. The solid is a rectangular block with a 3×4 base and height 2.

$$\iint_R 2\, dA = 3(4)2 = 24.$$

7) True, False:

a) $\displaystyle\iint_R xy\, dA = \iint_R x\, dA \iint_R y\, dA.$

False.

b) $\displaystyle\iint_R 6xy\, dA = 6\iint_R xy\, dA.$

True.

Section 16.2   Iterated Integrals

This section includes Fubini's Theorem – a method for calculating a double integral as two successive uses of the Fundamental Theorem of Calculus for single integrals. Just as we needed to hold a variable constant when performing partial differentiation, we will need to hold a variable constant in the "partial integration" in this section.

Concepts to Master

Iterated integrals; Fubini's Theorem (double integrals written as iterated integrals)

Summary and Focus Questions

Page 1026 (ET Page 990)

An *iterated integral* for the function f over the region $R = [a, b] \times [c, d]$ is defined as

$$\int_a^b \int_c^d f(x, y)\, dy\, dx = \int_a^b \left[\int_c^d f(x, y)\, dy \right] dx.$$

This is evaluated by first performing the partial integration of $f(x, y)$ with respect to y (holding x constant) and then integrate that result with respect to x, as in:

$$\int_1^2 \int_0^1 (2xy + x)dy\, dx = \int_1^2 \left[\int_0^1 (2xy + x)dy \right] dx = \int_1^2 \left[(xy^2 + xy) \Big]_{y=0}^1 \right] dx$$

$$= \int_1^2 (x + x - 0)dx = \int_1^2 2x\, dx = x^2 \Big]_{x=1}^2 = 4 - 1 = 3.$$

Fubini's Theorem: If f is continuous on $R = [a, b] \times [c, d]$, then

$$\iint_R f(x, y)\, dA = \int_a^b \int_c^d f(x, y)\, dy\, dx = \int_c^d \int_a^b f(x, y)\, dx\, dy.$$

1) Evaluate

   a) $\displaystyle \int_1^3 \int_0^2 (x^2 + 4y)dy\, dx$

$$\int_1^3 \left[\int_0^2 (x^2 + 4y)dy \right] dx$$

$$= \int_1^3 \left[(x^2 y + 2y^2) \Big]_{y=0}^2 \right] dx$$

$$= \int_1^3 (2x^2 + 8)dx$$

$$\left(\tfrac{2}{3}x^3 + 8x \right) \Big]_{x=1}^3 = \frac{100}{3}.$$

b) $\int_0^2 \int_0^3 8xy \, dx \, dy$

c) $\int_0^3 \int_0^2 8xy \, dy \, dx$

2) Find $\iint\limits_R (xy + e^x) dA$ where

$R = \{(x, y) \mid 0 \le x \le 1, 0 \le y \le 2\}$.

3) Write an iterated integral for

a) $\iint\limits_R (x^2 + y^2) dA$ where R is the

rectangle $0 \le x \le 4, 3 \le y \le 5$.

$\int_0^2 \left[\int_0^3 8xy \, dx \right] dy$

$= \int_0^2 \left(4x^2 y \right]_{x=0}^3 \right) dy = \int_0^2 36y \, dy$

$= 18y^2 \right]_{y=0}^2 \quad = 72 - 0 = 72.$

72. This is the iterated integral in question
1 b), written in reverse order.

$\iint\limits_R (xy + e^x) dA = \int_0^2 \int_0^1 (xy + e^x) dx \, dy$

$= \int_0^2 \left(\frac{x^2 y}{2} + e^x \right) \right]_{x=0}^1 dy$

$= \int_0^2 \left(\frac{y}{2} + e - 1 \right) dy$

$= \left(\frac{y^2}{4} + (e - 1)y \right) \right]_{y=0}^2$

$= 2e - 1.$

$\int_0^4 \int_3^5 (x^2 + y^2) \, dy \, dx$

$\left(\text{or } \int_3^5 \int_0^4 (x^2 + y^2) \, dx \, dy \right).$

b) the volume of the solid below

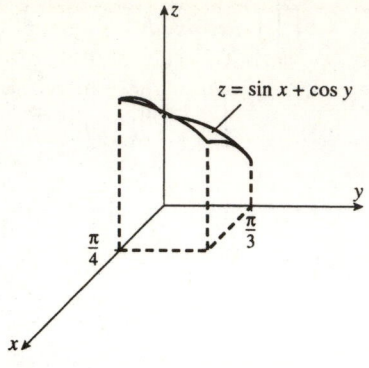

$$\int_0^{\frac{\pi}{4}}\int_0^{\frac{\pi}{3}}(\sin x + \cos y)dy\,dx$$

$$\left(\text{or } \int_0^{\pi/3}\int_0^{\pi/4}(\sin x + \cos y)dx\,dy\right).$$

4) True or False:
$$\int_0^3\int_0^4 4xy\,dy\,dx = \int_0^3\int_0^4 4xy\,dx\,dy$$

True, even though we canot use Fubini's Theorem, both integrals have value 144.

$$\int_0^3\int_0^4 4xy\,dy\,dx =$$
$$\int_0^3\left(2xy^2\Big]_{y=0}^4\right)dx$$
$$=\int_0^3 32x\,dx = 16x^2\Big]_{x=0}^3 = 144$$
$$\int_0^3\int_0^4 4xy\,dx\,dy =$$
$$\int_0^3\left(2x^2y\Big]_{x=0}^4\right)dy$$
$$=\int_0^3 32y\,dy = 16y^2\Big]_{y=0}^3 = 144$$

Section 16.3   **Double Integrals over General Regions**

This section extends the idea of a double integral to bounded regions other than rectangles. Fubini's Theorem may be applied to write them as iterated integrals.

Concepts to Master

A. Evaluate a double integral over a general region as an iterated integral

B. Change the order of integration of an iterated integral

C. Properties of double integrals

Summary and Focus Questions

Page 1032 (ET Page 996)

A. Let D be a region in the plane and $f(x, y)$ be continuous on D. If D may be described in either of the following ways, then $\iint\limits_D f(x, y)\, dA$ may be evaluated using an iterated integral:

Type I

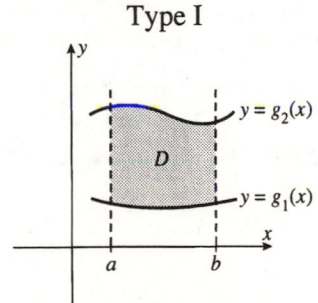

Type II

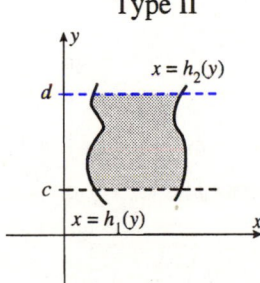

D described as
$a \leq x \leq b, g_1(x) \leq y \leq g_2(x).$

$$\iint\limits_D f(x, y)\, dA = \int_a^b \int_{g_1(x)}^{g_2(x)} f(x, y)\, dy\, dx$$

D described as
$c \leq y \leq d, h_1(y) \leq x \leq h_2(y).$

$$\iint\limits_D f(x, y)\, dA = \int_c^d \int_{h_1(x)}^{h_2(x)} f(x, y)\, dx\, dy$$

1) Write each as an iterated integral:

a) $\iint\limits_{D} xy^2 \, dA$, where D is the region below.

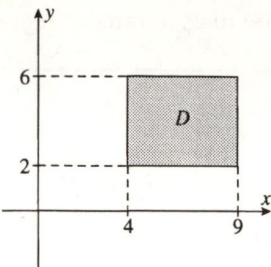

$$\int_{2}^{6} \int_{4}^{9} xy^2 \, dx \, dy \text{ or } \int_{4}^{9} \int_{2}^{6} xy^2 \, dy \, dx.$$

b) $\iint\limits_{D} e^{xy} \, dA$, where D is the region below.

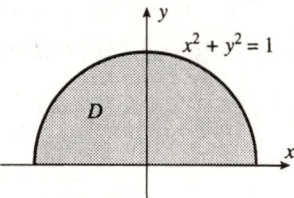

$x^2 + y^2 = 1$

The region may be described as Type I:
$-1 \le x \le 1, 0 \le y \le \sqrt{1 - x^2}.$

This integral is $\displaystyle\int_{-1}^{1} \int_{0}^{\sqrt{1 - x^2}} e^{xy} \, dy \, dx.$

c) $\iint\limits_{D} (x + y) \, dA$, where D is the region below.

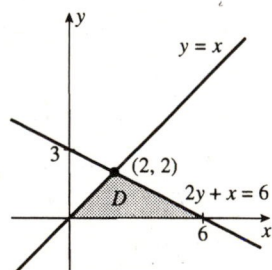

$y = x$

$(2, 2)$

$2y + x = 6$

Describing D as Type I would be difficult (the "top" function would be defined piecewise). D is easier to describe as Type II. $[h_1(y) = y$ and from $2y + x = 6$, $h_2(y) = 6 - 2y]$. The double integral is
$$\int_{0}^{2} \left[\int_{y}^{6 - 2y} (x + y) \, dx \right] dy.$$

2) Evaluate $\int_1^2 \int_y^{y^2} 4xy \, dx \, dy$.

$$\int_1^2 \int_y^{y^2} 4xy \, dx \, dy = \int_1^2 2x^2y \Big]_{x=y}^{y^2} dy$$

$$= \int_1^2 [2(y^2)^2 y - 2y^2 y] \, dy$$

$$= \int_1^2 (2y^5 - 2y^3) \, dy$$

$$= \left(\frac{y^6}{3} - \frac{y^4}{2} \right) \Big]_{y=1}^{2}$$

$$= \left(\frac{64}{3} - \frac{16}{2} \right) - \left(\frac{1}{3} - \frac{1}{2} \right) = \frac{27}{2}.$$

Page 1034 (ET Page 998)

B. An iterated integral corresponds to a double integral over a region D. If D may be described as both Type I and Type II it is possible to write the given form of the iterated integral in reverse form.

For example, $\int_0^3 \int_0^{6-2x} (x + y) \, dy \, dx$ corresponds to the region

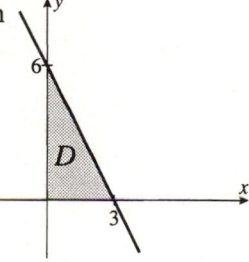

$D: 0 \le x \le 3, 0 \le y \le 6 - 2x$.

From $y = 6 - 2x$, $x = 3 - \frac{y}{2}$.

Hence D may be described $0 \le y \le 6, 0 \le x \le 3 - \frac{y}{2}$.

Thus $\int_0^3 \int_0^{6-2x} (x + y) \, dy \, dx = \int_0^6 \int_0^{3-y/2} (x + y) \, dx \, dy$.

3) Change the order of integration of

$$\int_0^4 \int_0^{\sqrt{4-x}} xy \, dy \, dx.$$

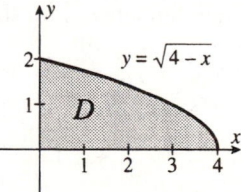

The iterated integral is $\iint_D xy \, dA$ where D is shown above. D may be rewritten as a Type II region:

$0 \le y \le 2$

$0 \le x \le 4 - y^2$

Thus the double integral is $\int_0^2 \int_0^{4-y^2} xy \, dx \, dy$.

C. If a region D may be described as the union of two nonoverlapping regions D_1 and D_2 then

Page 1037 (ET Page 1001)

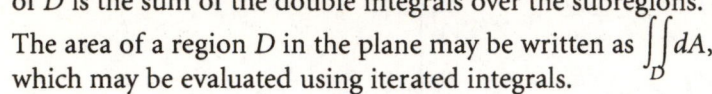

$$\iint_D f(x, y)dA = \iint_{D_1} f(x, y)dA + \iint_{D_2} f(x, y)dA$$

In particular, if D is neither Type I or Type II it may be possible to partition D into subregions of these types. The double integral of D is the sum of the double integrals over the subregions.

The area of a region D in the plane may be written as $\iint_D dA$, which may be evaluated using iterated integrals.

If f is integrable over D and $m \le f(x, y) \le M$ for all $(x, y) \in D$, then $m\, A(D) \le \iint_D f(x, y)dA \le M\, A(D)$, where $A(D)$ is the area of D.

4) Write an iterated integral for $\iint_D x^2 y\, dA$ where the region D is bounded by $y = 5x$, $y = x, x + y = 6$.

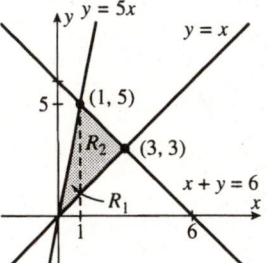

D may be divided into two Type I regions by the line $x = 1$. The integral is

$$\int_0^1 \int_x^{5x} x^2 y\, dy\, dx + \int_1^3 \int_x^{6-x} x^2 y\, dy\, dx.$$

5) Write an iterated integral for the shaded area.

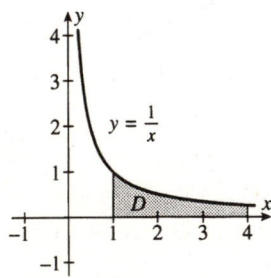

D may be described as $1 \le x \le 4$, $0 \le y \le \frac{1}{x}$. The area is $\int_1^4 \int_0^{1/x} dy\, dx$.

6) True or False:
$$\int_0^1 \int_3^5 x^2 y \, dy \, dx$$
$$= \int_0^1 \int_3^4 x^2 y \, dy \, dx + \int_0^1 \int_4^5 x^2 y \, dy \, dx.$$

True.

7) Let $D = \{(x, y) | 1 \le x \le y^2, 1 \le y \le 2\}$.

Using $1 \le x^2 y \le 32$ find bounds on
$$\iint_D x^2 y \, dA.$$

$m = 1$ and $M = 32$.

The area of D is

$$\iint_D dx \, dy = \int_1^2 \int_1^{y^2} dx \, dy$$

$$= \int_1^2 x \Big|_{x=1}^{y^2} dy$$

$$= \int_1^2 (y^2 - 1) dy = \left(\frac{y^3}{3} - y\right)\Big|_{y=1}^{2}$$

$$= \frac{4}{3}.$$

Thus

$$1\left(\frac{4}{3}\right) \le \iint_D x^2 y \, dA \le 32\left(\frac{4}{3}\right)$$

$$\frac{4}{3} \le \iint_D x^2 y \, dA \le \frac{128}{3}.$$

Section 16.4   Double Integrals in Polar Coordinates

This section shows you how to change a double integral into an iterated integral when the region of integration is described with polar coordinates.

Concepts to Master

Write a double integral as an iterated integral in polar coordinates

Summary and Focus Questions

Page 1041 (ET Page 1005)

Let D be a region in the plane and $f(x, y)$ be continuous on D. If D may be described using polar coordinates in either of the following ways, then $\iint_D f(x, y)dA$ may be written as an iterated integral in polar coordinates:

Type I

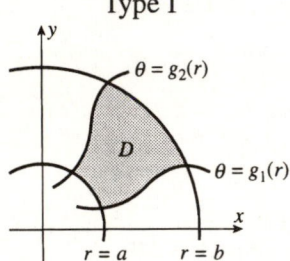

D described as

$a \le r \le b, g_1(r) \le \theta \le g_2(r)$.

$$\iint_D f(x, y)dA$$

$$= \int_a^b \int_{g_1(r)}^{g_2(r)} f(r \cos \theta, r \sin \theta)r \, d\theta \, dr$$

Type II

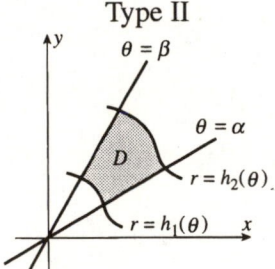

D described as

$\alpha \le \theta \le \beta, h_1(\theta) \le r \le h_2(\theta)$.

$$\iint_D f(x, y)dA$$

$$= \int_\alpha^\beta \int_{h_1(\theta)}^{h_2(\theta)} f(r \cos \theta, r \sin \theta)r \, dr \, d\theta$$

1) Rewrite, using polar coordinates:

a) $\iint_D x \, dA$ where D is the region below.

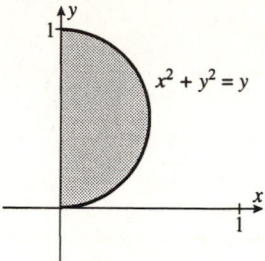

$$x^2 + y^2 = y$$

Converting $x^2 + y^2 = y$ to a polar equation gives $r^2 = r \sin \theta$, so $r = \sin \theta$.

The region D is Type II:

$$0 \le \theta \le \frac{\pi}{2}, 0 \le r \le \sin \theta.$$

Therefore the double integral is

$$\int_0^{\pi/2} \int_0^{\sin \theta} (r \cos \theta) r \, dr \, d\theta$$

(remember the "extra" r factor)

$$= \int_0^{\pi/2} \int_0^{\sin \theta} r^2 \cos \theta \, dr \, d\theta.$$

b) $\int_0^4 \int_0^x xy \, dy \, dx$

The corresponding region D is given by $0 \le x \le 4, 0 \le y \le x.$

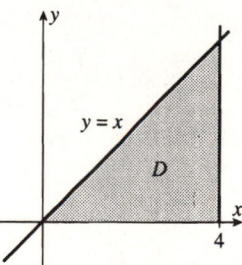

$$y = x$$

$$D$$

D may be described as a Type II region in polar coordinates:

$$0 \le \theta \le \frac{\pi}{4}, 0 \le r \le 4 \sec \theta$$

(the line $x = 4$ is $r \cos \theta = 4$, or $r = 4 \sec \theta$.)
Thus the double integral is

$$\int_0^{\pi/4} \int_0^{4 \sec \theta} (r \cos \theta)(r \sin \theta) r \, dr \, d\theta$$

$$= \int_0^{\pi/4} \int_0^{4 \sec \theta} r^3 \frac{\sin 2\theta}{2} \, dr \, d\theta.$$

2) Write a polar iterated integral for the volume under $z = x^2 + y^2$ and above the region D given here.

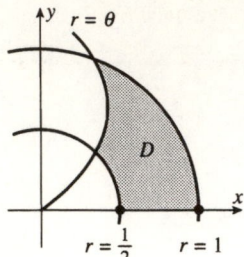

The region is Type I:
$\frac{1}{2} \le r \le 1, 0 \le \theta \le r$.
The volume under $z = x^2 + y^2$ is

$$\iint_D (x^2 + y^2)\,dA = \int_{1/2}^1 \int_0^r r^2 r\, d\theta\, dr$$
$$= \int_{1/2}^1 \int_0^r r^3\, d\theta\, dr.$$

Section 16.5   Applications of Double Integrals

This section includes some applications of double integrals for physical phenomenon and for probability distributions involving two variables.

Concepts to Master

A. Mass of a lamina; Center of mass; Moments of inertia
B. Probabilities involving joint density functions; Expected values

---

Summary and Focus Questions

Page
1046
(ET Page
1010)

A. Suppose a lamina (a flat plane area representing a distribution of matter) is described as a region D with density $\rho(x, y)$ at each point in D. Let ρ be continuous.

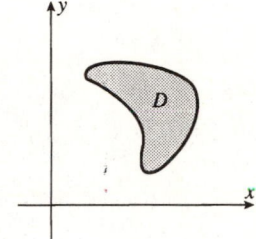

The *mass* of the lamina is $m = \iint\limits_D \rho(x, y)dA.$

The *moment of mass* with respect to the x-axis is $M_x = \iint\limits_D y\rho(x, y)dA.$

The *moment of mass* with respect to the y-axis is $M_y = \iint\limits_D x\rho(x, y)dA.$

The *center of mass* is $(\bar{x}, \bar{y})$, where $\bar{x} = \dfrac{M_y}{m}$ and $\bar{y} = \dfrac{M_x}{m}.$

Moments of inertia of a lamina measure the tendency to rotate about a line or point.
The *moment of inertia about the x-axis* is $I_x = \iint\limits_D y^2\rho(x, y) \, dA.$

The *moment of inertia about the y-axis* is $I_y = \iint\limits_D x^2\rho(x, y) \, dA.$

The *moment of inertia about the origin* is $I_0 = I_x + I_y.$

1) Find the mass of the lamina below whose density at each point (x, y) is $72xy$.

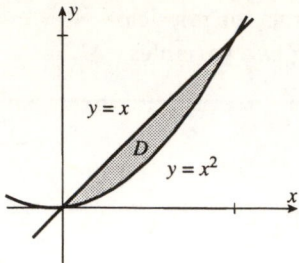

$$m = \iint_D 72xy \, dA = \int_0^1 \int_{x^2}^x 72xy \, dy \, dx$$

$$= \int_0^1 36xy^2 \Big]_{y=x^2}^x dx$$

$$= \int_0^1 (36x^3 - 36x^5)\,dx$$

$$= (9x^4 - 6x^6)\Big]_{x=0}^1 = 3.$$

2) Find the y-coordinate of the center of mass of the lamina in question 1.

$$M_x = \iint_D y(72xy)\,dA = \iint_D 72xy^2 \, dA$$

$$= \int_0^1 \int_{x^2}^x 72xy^2 \, dy \, dx$$

$$= \int_0^1 (24x^4 - 24x^7)\,dx = \tfrac{9}{5}.$$

Thus $\bar{y} = \dfrac{M_x}{m} = \dfrac{9/5}{3} = \dfrac{3}{5}.$

3) Find an iterated integral for the moment of inertia about the y-axis of the lamina in question 1.

$$I_y = \iint_D x^2(72xy)\,dA = \int_0^1 \int_{x^2}^x 72x^3 y \, dy \, dx.$$

B. A *joint density function* f of variables x and y is a function such that the probability that (x, y) lies in a region D is

Page
1050
(ET Page
1014)

$$P((x, y) \in D) = \iint_D f(x, y)\,dA$$

and has these properties

$$f(x, y) \geq 0$$

$$\iint_{\mathbb{R}^2} f(x, y)\,dA = 1$$

If x and y are *independent* then there are functions f_1 and f_2 so that

$$f(x, y) = (f_1(x))(f_2(y)).$$

The *X-mean* for a joint density function f is $\mu_1 = \iint_{\mathbb{R}^2} x f(x, y)\,dA$.

The *Y-mean* for a joint density function f is $\mu_2 = \iint_{\mathbb{R}^2} y f(x, y)\,dA$.

4) Let $f(x, y) = \begin{cases} \dfrac{e^{x+y}}{(e-1)^2}, & 0 \leq x \leq 1, \\ & 0 \leq y \leq 1, \\ 0 & \text{otherwise} \end{cases}$

a) Show that f is a joint probability function.

Clearly, $f(x, y) \geq 0$ for all x and y.

$$\iint_{\mathbb{R}^2} f(x, y)\,dA = \int_0^1 \int_0^1 \frac{e^{x+y}}{(e-1)^2}\,dy\,dx$$

$$= \int_0^1 \left(\frac{e^{x+y}}{(e-1)^2} \Big]_0^1 \right) dx = \int_0^1 \frac{e^{x+1} - e^x}{(e-1)^2}\,dx$$

$$= \frac{e^{x+1} - e^x}{(e-1)^2} \Big]_0^1 = \frac{e^2 - e - (e-1)}{(e-1)^2}$$

$$= 1.$$

b) Are x and y independent?

Yes, because we may write $f(x, y)$ as

$$\frac{e^{x+y}}{(e-1)^2} = \left(\frac{e^x}{e-1} \right)\left(\frac{e^y}{e-1} \right).$$

c) Find $P\left(0 \leq x \leq \frac{1}{2}, 0 \leq y \leq \frac{1}{3}\right)$.

$$\int_0^{1/2} \int_0^{1/3} \frac{e^{x+y}}{(e-1)^2}\,dy\,dx = \int_0^1 \left(\frac{e^{x+y}}{(e-1)^2} \Big]_0^{1/3} \right) dx$$

$$= \int_0^1 \frac{e^{x+1/3} - e^x}{(e-1)^2}\,dx = \frac{e^{x+1/3} - e^x}{(e-1)^2} \Big]_0^{1/2}\,dx$$

$$= \frac{e^{5/6} - e^{1/2} - (e^{1/3} - e^0)}{(e-1)^2}$$

$$= \frac{e^{5/6} - e^{1/2} - e^{1/3} + 1}{(e-1)^2} \approx 0.087.$$

d) Find the X-mean.

$$\iint_{\mathbb{R}^2} x f(x, y)dA = \int_0^1 \int_0^1 \frac{x\, e^{x+y}}{(e-1)^2}dy\, dx$$

$$= \int_0^1 \left(\frac{x\, e^{x+y}}{(e-1)^2}\right]_0^1\right)dx = \int_0^1 \frac{x(e^{x+1} - e^x)}{(e-1)^2}\, dx$$

$$= \int_0^1 \frac{x e^x(e-1)}{(e-1)^2}dx$$

$$= \int_0^1 \frac{x e^x}{e-1}\, dx = \text{(integration by parts)}$$

$$= \frac{x e^x - e^x}{e-1}\bigg]_0^1 = \frac{e^1 - e^1}{e-1} - \frac{0-1}{e-1}$$

$$= \frac{1}{e-1} \approx 0.582.$$

Section 16.6   Surface Area

This section computes the area of a surface using a double integral.

Concepts to Master

Area of a surface given by the equation $z = f(x, y)$

Summary and Focus Questions

Page 1056 (ET Page 1020)

The area of the surface $z = f(x, y)$ where f has continuous partial derivatives over a region D is

$$\iint_D \sqrt{[f_x(x, y)]^2 + [f_y(x, y)]^2 + 1} \, dA.$$

1) Write an iterated integral for the surface area of the paraboloid $z = x^2 + 2y^2$ above the triangular region D:

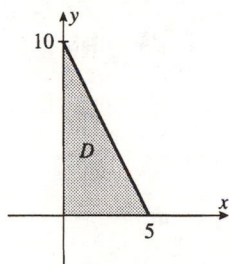

$f_x = 2x, f_y = 4y.$

The region D is $0 \le x \le 5$,
$0 \le y \le 10 - 2x$.

The surface area is

$$\iint_D \sqrt{(2x)^2 + (4y)^2 + 1} \, dA$$

$$= \int_0^5 \int_0^{10 - 2x} \sqrt{4x^2 + 16y^2 + 1} \, dy \, dx.$$

2) Find the surface area of the top portion of the sphere $x^2 + y^2 + z^2 = 4$ inside the cylinder $x^2 + y = 1$.

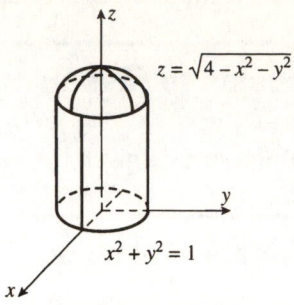

For $z = \sqrt{4 - x^2 - y^2}$,

$$\frac{\partial z}{\partial x} = \frac{-x}{\sqrt{4 - x^2 - y^2}} \text{ and } \frac{\partial z}{\partial y} = \frac{-y}{\sqrt{4 - x^2 - y^2}}.$$

Thus $1 + \left(\frac{\partial z}{\partial x}\right)^2 + \left(\frac{\partial z}{\partial y}\right)^2 = $

$$1 + \frac{x^2}{4 - x^2 - y^2} + \frac{y^2}{4 - x^2 - y^2} = \frac{4}{4 - x^2 - y^2}$$

The surface area is

$$\iint_D \frac{2}{\sqrt{4 - x^2 - y^2}} \, dA, \text{ where } D \text{ is the area}$$

inside the circle $x^2 + y^2 = 1$. In polar

coordinates, this is

$$\int_0^{2\pi} \int_0^1 \frac{2}{\sqrt{4 - r^2}} r \, dr \, d\theta$$

$$= \int_0^{2\pi} \left(-2\sqrt{4 - r^2} \Big|_0^1 \right) d\theta$$

$$= \int_0^{2\pi} (-2\sqrt{3} - (-2)(2)) \, d\theta$$

$$= \int_0^{2\pi} (4 - 2\sqrt{3}) \, d\theta$$

$$= 2\pi(4 - 2\sqrt{3}) = 8\pi - 4\pi\sqrt{3}.$$

Section 16.7   Triple Integrals

This section extends all the concepts of double integrals to functions of three variables.

Concepts to Master

A. Triple Integrals; Fubini's Theorem for triple integrals

B. Applications of the triple integral as volume, mass, center of mass, and moment of inertia

---

Summary and Focus Questions

Page 1059 (ET Page 1023)

A. A triple integral of a function $f(x, y, z)$ over a region E in a three-dimensional space $\iiint_E f(x, y, z)dV$ is defined in the usual way as a limit of Riemann sums:

$$\lim_{l, m, n \to \infty} \sum_{i=1}^{l} \sum_{j=1}^{m} \sum_{k=1}^{n} f(x_i^*, y_j^*, z_k^*)\Delta V$$

where we partition E into sub-boxes and (x_i^*, y_j^*, z_k^*) is chosen from the i-j-k-th sub-box whose volume is $\Delta V = \Delta x \, \Delta y \, \Delta z$.

Fubini's Theorem: A triple integral may be evaluated as one of six possible iterated integrals. For example, if E can be described by $a \le x \le b$, $g_1(x) \le y \le g_2(x)$, $h_1(x, y) \le z \le h_2(x, y)$ then

$$\iiint_E f(x, y, z)dV = \int_a^b \int_{g_1(x)}^{g_2(x)} \int_{h_1(x, y)}^{h_2(x, y)} f(x, y, z)dz \, dy \, dx.$$

As with an iterated double integral, this is evaluated "from the inside out;" that is, integrate with respect to z and substitute the z-limits of integration, then with respect to y and substitute the y-limits, and finally with respect to x and substitute the x-limits.

1) Evaluate $\displaystyle\int_1^2 \int_1^y \int_0^{x+y} 12x \, dz \, dx \, dy$.

$$\int_1^2 \int_1^y \int_0^{x+y} 12x \, dz \, dx \, dy$$

$$= \int_1^2 \int_1^y \left(12xz\Big]_{z=0}^{x+y}\right) dx \, dy$$

$$= \int_1^2 \int_1^y (12x^2 + 12xy) dx \, dy$$

$$= \int_1^2 (4x^3 + 6x^2y)\Big]_{x=1}^{y} dy$$

$$= \int_1^2 (4y^3 + 6y^3) - (4 + 6y) dy$$

$$= \int_1^2 (10y^3 - 6y - 4) dy$$

$$= \left(\frac{10}{4}y^4 - 3y^2 - 4y\right)\Big]_{y=1}^{2} = \frac{49}{2}.$$

2) $\displaystyle\int_1^4 \int_2^5 \int_3^7 xyz \, dx \, dz \, dy = \int_?^? \int_?^? \int_?^? xyz \, dz \, dy \, dx$

$$\int_3^7 \int_1^4 \int_2^5 xyz \, dz \, dy \, dx.$$

3) Write an iterated integral for $\displaystyle\iiint_E x \, dV$, where E is the region cut off in the first octant by the plane $x + 2y + 4z = 8$.

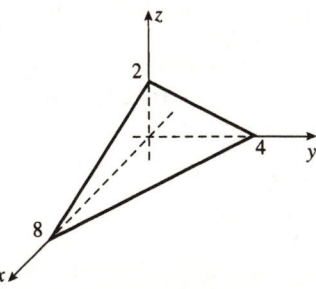

There are six possible iterated integrals as answers, each depending upon how the solid E is represented. We write E as

$0 \le z \le 2, 0 \le y \le 4 - 2z,$
$0 \le x \le 8 - 2y - 4z.$
Thus

$$\iiint_E x \, dV = \int_0^2 \int_0^{4-2z} \int_0^{8-2y-4z} x \, dx \, dy \, dz.$$

4) Write $\displaystyle\iiint_E (x + y)\,dV$ as an iterated integral where E is the solid ball of radius 2 about the origin.

The region is inside the sphere $x^2 + y^2 + z^2 = 4$. E may be described by
$-2 \le x \le 2$,
$-\sqrt{4 - x^2} \le y \le \sqrt{4 - x^2}$,
$-\sqrt{4 - x^2 - y^2} \le z \le \sqrt{4 - x^2 - y^2}$.
Thus $\displaystyle\iiint_E (x + y)\,dV$

$$= \int_{-2}^{2} \int_{\sqrt{4-x^2}}^{-\sqrt{4-x^2}} \int_{\sqrt{4-x^2-y^2}}^{-\sqrt{4-x^2-y^2}} (x+y)\,dz\,dy\,dx.$$

5) Write $\displaystyle\int_0^1 \int_0^y \int_0^{1-y} y^2 \,dx\,dz\,dy$ as an iterated integral in the form $\displaystyle\iiint y^2\,dz\,dy\,dx$.

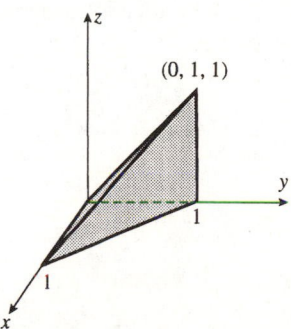

The space for the given iterated integral is the shaded volume above, which may be written as
$0 \le y \le 1, 0 \le z \le y, 0 \le x \le 1 - y$.

For $\displaystyle\iiint y^2\,dz\,dy\,dx$, the region is $0 \le x \le 1$,
$0 \le y \le 1 - x, 0 \le z \le y$.

Thus the desired iterated integral is
$$\int_0^1 \int_0^{1-x} \int_0^y y^2 \,dz\,dy\,dx.$$

Page
1064
(ET Page
1028)

B. The volume of a solid E is $\iiint\limits_E dV$. If a solid E has a continuous density $\rho(x, y, z)$, the *mass of the solid, m,* is $\iiint\limits_E \rho(x, y, z)dV$.

The three *moments* are:

about *xy*-plane	about *xz*-plane	about *yz*-plane
$M_{xy} = \iiint\limits_E z\rho(x, y, z)dV$	$M_{xz} = \iiint\limits_E y\rho(x, y, z)dV$	$M_{yz} = \iiint\limits_E x\rho(x, y, z)dV$

The *moments of inertia* about the axes are:

$$I_x = \iiint\limits_E (y^2 + z^2)\rho(x, y, z)dV$$

$$I_y = \iiint\limits_E (x^2 + z^2)\rho(x, y, z)dV$$

$$I_z = \iiint\limits_E (x^2 + y^2)\rho(x, y, z)dV$$

6) Find a triple iterated integral for the volume inside the cone $z^2 = x^2 + y^2$ bounded by $z = 0, z = 8$.

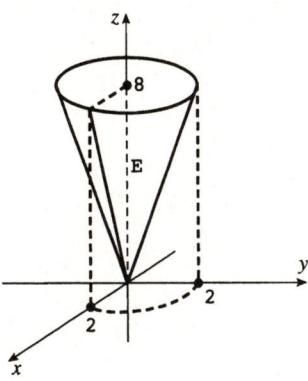

The solid E may be described as

$0 \le z \le 8, -\frac{z}{4} \le y \le \frac{z}{4},$
$-\sqrt{z^2 - y^2} \le x \le \sqrt{z^2 - y^2}.$

The volume is $\int_0^8 \int_{-z/4}^{z/4} \int_{\sqrt{z^2 - y^2}}^{-\sqrt{z^2 - y^2}} dx\, dy\, dz.$

7) A solid has density xyz and is described as bounded by the cylinder $x = z^2$ and the planes $y = 0, x = 0, y = 5, z = 2$.

  a) Find an expression for the z-coordinate of the center mass.

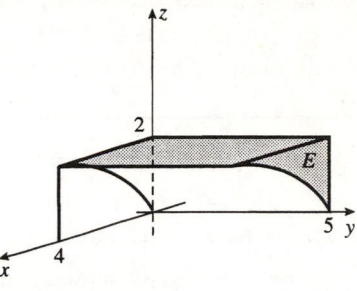

The solid E may be described as $0 \le y \le 5$, $0 \le z \le 2, 0 \le x \le z^2$.

The mass $m = \iiint_E xyz \, dV$

$$= \int_0^5 \int_0^2 \int_0^{z^2} xyz \, dx \, dz \, dy.$$

$$M_{xy} = \iiint_E z(xyz)dV$$

$$= \int_0^5 \int_0^2 \int_0^{z^2} xyz^2 \, dx \, dz \, dy.$$

$$\bar{z} = \frac{M_{xy}}{m}.$$

  b) Find an expression for the moment of inertia about the x-axis.

$$I_x = \iiint_E (y^2 + z^2)xyz \, dV$$

$$= \int_0^5 \int_0^2 \int_0^{z^2} (xy^3z + xyz^3)dx \, dz \, dy.$$

Section 16.8   Triple Integrals in Cylindrical and Spherical Coordinates

This section extends the notion of polar coordinates for double integrals to cylindrical coordinates for triple integrals. It also discusses spherical coordinates for triple integrals. As before, the advantage is that some integrals are easier to evaluate in one coordinate system over the others.

Concepts to Master

A. Triple integrals as iterated integrals in cylindrical coordinates

B. Triple integrals as iterated integrals in spherical coordinates

Summary and Focus Questions

Page 1069 (ET Page 1033)

A. If f is continuous on a solid region E and E may be described in cylindrical coordinates by $\alpha \le \theta \le \beta$, $h_1(\theta) \le r \le h_2(\theta)$ $\phi_1(x, y) \le z \le \phi_2(x, y)$, then

$$\iiint_E f(x, y, z)\,dV = \int_\alpha^\beta \int_{h_1(\theta)}^{h_2(\theta)} \int_{\phi_1(r \cos \theta, r \sin \theta)}^{\phi_2(r \cos \theta, r \sin \theta)} f(r \cos \theta, r \sin \theta, z)r\,dz\,dr\,d\theta$$

1) Evaluate $\iiint_E 2\,dV$ where E is the solid between the paraboloids

$$z = 4x^2 + 4y^2$$
$$z = 80 - x^2 - y^2.$$

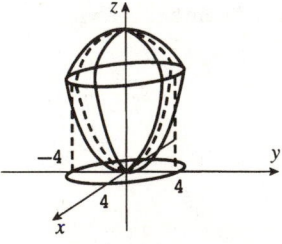

The paraboloids intersect in the circle $4x^2 + 4y^2 = 80 - x^2 - y^2$.

This simplifies to $x^2 + y^2 = 16$.
E may be described in cylindrical coordinates as $0 \le \theta \le 2\pi$, $0 \le r \le 4$, $4x^2 + 4y^2 \le z \le 80 - x^2 - y^2$.

Thus $4r^2 \le z \le 80 - r^2$.

$$\iiint\limits_{E} 2 \, dV = \int_0^{2\pi} \int_0^4 \int_{4r^2}^{80-r^2} 2r \, dz \, dr \, d\theta$$

$$= \int_0^{2\pi} \int_0^4 2zr \Big]_{z=4r^2}^{80-r^2} dr \, d\theta$$

$$= \int_0^{2\pi} \int_0^4 (160r - 10r^3) dr \, d\theta$$

$$= \int_0^{2\pi} \left(80r^2 - \frac{5}{2}r^4\right)\Big]_{r=0}^{4} d\theta$$

$$= \int_0^{2\pi} 640 \, d\theta = 640\theta \Big]_{\theta=0}^{2\pi} = 1280\pi.$$

B. If f is continuous on a solid region E and E may be described in spherical coordinates by $\alpha \leq \theta \leq \beta, c \leq \phi \leq d, g_1(\theta, \phi) \leq \rho \leq g_2(\theta, \phi)$, then

$$\iiint\limits_{E} f(x, y, z) dV$$

$$= \int_c^d \int_\alpha^\beta \int_{g_1(\theta, \phi)}^{g_2(\theta, \phi)} f(\rho \sin \phi \cos \theta, \rho \sin \phi \sin \theta, \rho \cos \phi)\rho^2 \sin \phi \, d\rho \, d\theta \, d\phi.$$

Page 1071 (ET Page 1035)

2) Write as an iterated integral in spherical coordinates: $\iiint\limits_{E} (10 - x^2 - y^2 - z^2) dV$, where E is the top half of the sphere $x^2 + y^2 + z^2 = 9$.

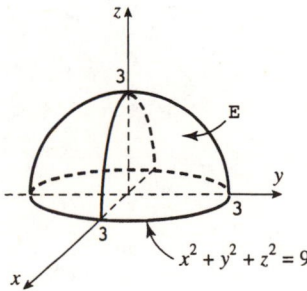

E may be described in spherical coordinates as $0 \leq \rho \leq 3, 0 \leq \theta \leq 2\pi, 0 \leq \phi \leq \frac{\pi}{2}$.

$$\iiint\limits_{E} (10 - x^2 - y^2 - z^2) dV$$

$$= \int_0^{\pi/2} \int_0^{2\pi} \int_0^3 (10 - \rho^2)\rho^2 \sin \phi \, d\rho \, d\theta \, d\phi.$$

3) Which system, spherical or cylindrical, would be more appropriate to evaluate $\iiint_E x\,dV$ where E is the solid between the cone $z^2 = x^2 + y^2$ and the plane $z = 4$?

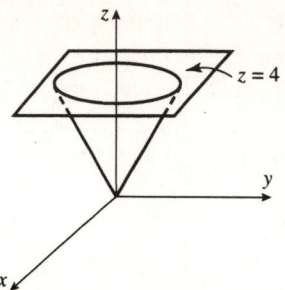

Cylindrical, since E may be written as $0 \leq \theta \leq 2\pi, 0 \leq r \leq 2, \sqrt{x^2 + y^2} \leq z \leq 4.$

Section 16.9   Change of Variables in Multiple Integrals

Change of variable for a single variable integral involved determining an intermediate function $u = g(x)$ and corresponding $du = g'(x)dx$. The function $u = g(x)$ may be thought of as a transformation of the reals (x is "transformed" to $g(x)$). In higher dimensions, there are several intermediate variables involved. The analog of du is found from a matrix whose determinant is called a Jacobian.

Concepts to Master

A. C^1 transformations; Jacobians in 2 and 3 dimensions

B. Change of variable for double and triple integrals

Summary and Focus Questions

A. A *transformation* in the plane is a function T from the uv-plane to the xy-plane

$$T(u, v) = (x, y), \text{ where } x = g(u, v) \text{ and } y = h(u, v).$$

Page 1077 (ET Page 1041)

T is C^1 means g and h have continuous first derivatives.

Similarly, a transformation in space is a function T from the uvw-space to xyz-space.

For a C^1 transformation T from uv to xy, the *Jacobian of T* is the determinant

$$\frac{\partial(x, y)}{\partial(u, v)} = \begin{vmatrix} \dfrac{\partial x}{\partial u} & \dfrac{\partial x}{\partial v} \\ \dfrac{\partial y}{\partial u} & \dfrac{\partial y}{\partial v} \end{vmatrix}.$$

In three dimensions the Jacobian is

$$\frac{\partial(x, y, z)}{\partial(u, v, w)} = \begin{vmatrix} \dfrac{\partial x}{\partial u} & \dfrac{\partial x}{\partial v} & \dfrac{\partial x}{\partial w} \\ \dfrac{\partial y}{\partial u} & \dfrac{\partial y}{\partial v} & \dfrac{\partial y}{\partial w} \\ \dfrac{\partial z}{\partial u} & \dfrac{\partial z}{\partial v} & \dfrac{\partial z}{\partial w} \end{vmatrix}.$$

1) Find the Jacobian of each transformation:

    a) $x = 2u + v$
          $y = u - 2v$

$$\frac{\partial(x, y)}{\partial(u, v)} = \begin{vmatrix} 2 & 1 \\ 1 & -2 \end{vmatrix} = (2)(-2) - (1)(1) = -5.$$

    b) $x = uv$
          $y = u + v$

$$\frac{\partial(x, y)}{\partial(u, v)} = \begin{vmatrix} v & u \\ 1 & 1 \end{vmatrix} = v - u.$$

c) $x = 2u + v$
$y = uw$
$z = u^2 + v^2 + w^2$

$$\frac{\partial(x, y, z)}{\partial(u, v, w)} = \begin{vmatrix} 2 & 1 & 0 \\ w & 0 & u \\ 2u & 2v & 2w \end{vmatrix}$$

$$= 2\begin{vmatrix} 0 & u \\ 2v & 2w \end{vmatrix} - 1\begin{vmatrix} w & u \\ 2u & 2w \end{vmatrix} + 0\begin{vmatrix} w & 0 \\ 2u & 2v \end{vmatrix}$$

$$= 2(-2uv) - 1(2w^2 - 2u^2) + 0$$

$$= 2u^2 - 4uv - 2w^2.$$

2) Find the region S that maps to the region R under the transformation T:

$$x = \tfrac{2}{3}(v - u)$$

$$y = \tfrac{1}{3}(u + 2v).$$

R is the shaded region:

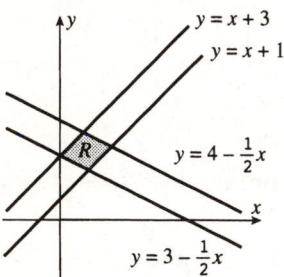

We find T^{-1} by solving for u and v. Multiplying each of $x = \tfrac{2}{3}(v - u)$ and $y = \tfrac{1}{3}(u + 2v)$ by 3 gives

$$3x = 2v - 2u$$
$$3y = 2v + u.$$

Subtracting the equations gives

$3x - 3y = -3u$. Thus $u = y - x$.

Then $3x = 2v - 2(y - x)$, so $v = \tfrac{1}{2}x + y$.

The four lines are transformed by T^{-1} from the xy-plane to the uv-plane as follows:

xy	uv
$y = 4 - \tfrac{1}{2}x$	$v = 4$
$y = 3 - \tfrac{1}{2}x$	$v = 3$
$y = x + 3$	$u = 3$
$y = x + 1$	$u = 1$

Thus $S = \{(u, v) \mid 1 \le u \le 3, 3 \le v \le 4\} = [1, 3] \times [3, 4]$ and S is mapped to R by T.

Page 1080 (ET Page 1044)

B. Let T be a one-to-one C^1 transformation that maps a Type I or II region S in the uv-plane onto a Type I or II region R in the xy-plane. If T has a nonzero Jacobian and f is continuous, then

$$\iint_R f(x, y)dx\, dy = \iint_S f(x(u, v), y(u, v))\left|\frac{\partial(x, y)}{\partial(u, v)}\right|du\, dv.$$

A similar result holds for transformations in three dimensions.

3) Use question 2 to change the variable in $\iint_R (6x + 3y)dA$ and evaluate.

With the transformation T:

$x = \frac{2}{3}(v - u),\ y = \frac{1}{3}(u + 2v)$

$T(S) = R$ where $S = [1, 3] \times [3, 4]$.

The Jacobian of T is

$$\begin{vmatrix} \frac{\partial x}{\partial u} & \frac{\partial x}{\partial v} \\ \frac{\partial y}{\partial u} & \frac{\partial y}{\partial v} \end{vmatrix} = \begin{vmatrix} -\frac{2}{3} & \frac{2}{3} \\ \frac{1}{3} & \frac{2}{3} \end{vmatrix} = -\frac{4}{9} - \frac{2}{9} = -\frac{2}{3}$$

Thus $\iint_R (6x + 3y)dA$

$$= \iint_S \left(6\left[\frac{2}{3}(v - u)\right]\right.$$

$$\left. + 3\left[\frac{1}{3}(u + 2v)\right]\right)\left|-\frac{2}{3}\right|du\, dv$$

$$= \iint_S (4v - 2u)du\, dv$$

$$= \int_3^4 \int_1^3 (4v - 2u)du\, dv$$

$$= \int_3^4 (4uv - u^2)\Big]_{u=1}^3 dv$$

$$= \int_3^4 (8v - 8)dv$$

$$= (4v^2 - 8v)\Big]_{v=3}^4 = 20.$$

4) Use a change of variable to evaluate $\iint_R (x^2 - y^2)dx\,dy$ where R is the shaded region.

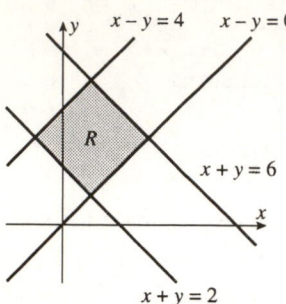

The region is bounded by lines whose equations include the expressions $x + y$ and $x - y$.

We let $u = x + y$, $v = x - y$.

Then $2 \le u \le 6$, $0 \le v \le 4$.

We must write x, y in terms of u, v:

$$u = x + y$$
$$v = x - y$$

$u + v = 2x$, so $x = \dfrac{u + v}{2}$.

Thus $u = \dfrac{u + v}{2} + y$, so $y = \dfrac{u - v}{2}$.

The integrand $f(x, y) = x^2 - y^2$

$$= \left(\frac{u + v}{2}\right)^2 - \left(\frac{u - v}{2}\right)^2 = uv.$$

Finally, the Jacobian is

$$\begin{vmatrix} \dfrac{\partial x}{\partial u} & \dfrac{\partial x}{\partial v} \\ \dfrac{\partial y}{\partial u} & \dfrac{\partial y}{\partial v} \end{vmatrix} = \begin{vmatrix} \dfrac{1}{2} & \dfrac{1}{2} \\ \dfrac{1}{2} & -\dfrac{1}{2} \end{vmatrix} = -\frac{1}{4} - \frac{1}{4} = -\frac{1}{2}.$$

Thus

$$\iint_R (x^2 - y^2)dx\,dy = \int_2^6 \int_0^4 uv \left|-\frac{1}{2}\right| dv\,du$$

$$= \int_2^6 \int_0^4 \frac{uv}{2}\,dv\,du = \int_2^6 \frac{uv^2}{4}\bigg]_{v=0}^{4} du$$

$$= \int_2^6 4u\,du = 2u^2 \bigg]_{u=2}^{6} = 64.$$

Technology Plus for Chapter 16

1) Use a spreadsheet or a calculator to estimate $\int_2^4 \int_1^5 (x + xy + y)\,dA$ using the Midpoint Rule with $\Delta x = 0.5$ and $\Delta y = 1$.

Let $f(x, y) = x + xy + y$.

$\Delta x = 0.5$

$\Delta y = 1$

$\Delta A = 0.5$

	y			
	1.5	2.5	3.5	4.5
2.25	7.125	10.375	13.625	16.875
x 2.75	8.375	12.125	15.875	19.625
3.25	9.625	13.875	18.125	22.375
3.75	10.875	15.625	20.375	25.125

$$\sum_{i=1}^{4} \sum_{j=1}^{4} f(x_{ij}, y_{ij}) = 240$$

$$\sum_{i=1}^{4} \sum_{j=1}^{4} f(x_{ij}, y_{ij})\Delta A = 120.$$

2) Use a CAS to draw the solid bounded by $0 \le x \le \frac{\pi}{2}, 0 \le y \le x, 0 \le z \le \sin x \cos y$. Then find the volume of the solid.

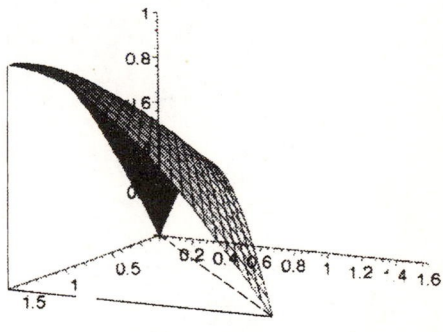

The volume is $\int_0^{\pi/2} \int_0^x \sin x \cos y \, dy \, dx$

$$= \int_0^{\pi/2} (\sin x \sin y) \Big]_{y=0}^{x} dx$$

$$= \int_0^{\pi/2} \sin^2 x \, dx$$

$$= \left(\frac{x}{2} - \frac{\sin x \cos x}{2}\right)\Big]^{\pi/2} = \frac{\pi}{4}.$$

3) Use a CAS to sketch the surface
$z = x + y - x^2 - y^2, 0 \leq x \leq 1, 0 \leq y \leq 1$,
and find its area to three decimal places.

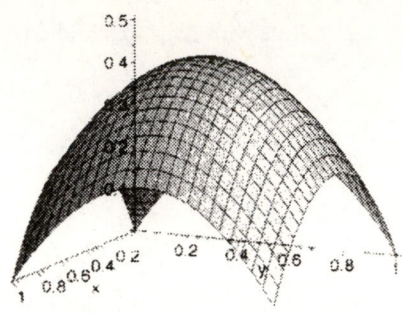

$f(x, y) = x + y - x^2 - y^2$
$f_x = 1 - 2x, f_y = 1 - 2y.$

The surface area is

$$\iint_D \sqrt{(1 - 2x)^2 + (1 - 2y)^2 + 1} \, dA,$$

where D is the $[0, 1] \times [0, 1]$ square.

The integral is

$$\int_0^1 \int_0^1 \sqrt{3 - 4x + 4x^2 - 4y + 4y^2} \, dy \, dx$$

which, to three decimals, is 1.281.

Chapter 17 — Vector Calculus

"BUT WE JUST DON'T HAVE THE TECHNOLOGY TO CARRY IT OUT."

Cartoons courtesy of Sidney Harris. Used by permission.

Section 17.1   Vector Fields

This section describes functions from R^2 to R^2, or from R^3 to R^3. Such functions are known as vector fields because we think of the domain elements as points in a set (a field of points in a plane or three dimensions) and range elements as two- or three-dimensional vectors associated with those points. A conservative vector field is one that is the gradient of a function.

Concepts to Master

Vector fields in two and three dimensions; Conservative vector fields

Summary and Focus Questions

Page
1092
(ET Page
1056)*

A vector field whose domain is a set D in R^2 (or R^3) is a function **F** that assigns to each point (x, y) in D a vector $\mathbf{F}(x, y)$. We think of domain elements as points and range elements as vectors. A two-dimensional vector field may be represented by drawing the plane, choosing representative points and sketching the vectors associated with those points.

When points in D are thought of as vectors, $\mathbf{x} = (x, y)$, we write $\mathbf{F}(\mathbf{x})$. Also, the vector $\mathbf{F}(x, y)$ may be written in terms of its *scalar field* component functions:

$$\mathbf{F}(x, y) = P(x, y)\mathbf{i} + Q(x, y)\mathbf{j}$$

For example, let $\mathbf{F}(x, y) = x\mathbf{i} + 2\mathbf{j}$. The component functions are $P(x, y) = x$ and $Q(x, y) = 2$. Here is a sketch of **F** with four domain points $(0, 0)$, $(1, -4)$, $(3, 1)$, and $(-1, -1)$ and their associated vectors:

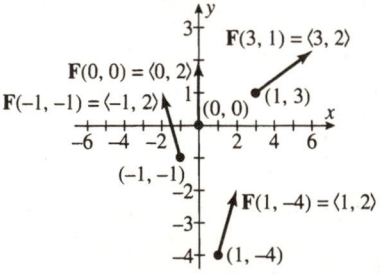

The concepts are similar in three dimensions. For example,
$\mathbf{F}(x, y, z) = (y + z)\mathbf{i} + (x + z)\mathbf{j} + (x + y)\mathbf{k}$ is a three dimensional vector field with component functions $P(x, y, z) = y + z$, $Q(x, y, z) = x + z$, $R(x, y, z) = x + y$.

*Remember, when using the Early Transcendentals book, use the page in parentheses!

For a scalar function $z = f(x, y)$, the gradient ∇f, is a vector field. In two dimensions
$$\nabla f(x, y) = f_x(x, y)\mathbf{i} + f_y(x, y)\mathbf{j}.$$

A vector field $\mathbf{F}$ is *conservative* means $\mathbf{F} = \nabla f$ for some scalar function f called the *potential field of $\mathbf{F}$*. For example, $\mathbf{F}(x, y) = 2xy\mathbf{i} + x^2\mathbf{j}$ is conservative because $\mathbf{F} = \nabla f$, where $f(x, y) = x^2 y$ (Note: $f_x = 2xy$ and $f_y = x^2$, the components of $\mathbf{F}$). We will have a means for checking for conservativeness in Section 17.3.

1) What are the scalar field components of

   a) $F(x, y) = \cos x\mathbf{i} + \sin xy\mathbf{j}$

                                                 $P(x, y) = \cos x$ and $Q(x, y) = \sin xy$

   b) $F(x, y, z) = (x^2 + y)\mathbf{i} + (y^2 + z)\mathbf{k}$

                                                 $P(x, y, z) = x^2 + y$
                                                 $Q(x, y, z) = 0$
                                                 $R(x, y, z) = y^2 + z$

2) For the vector field $\mathbf{F}(x, y) = |x|\mathbf{i} + |y|\mathbf{j}$, sketch the images of $(0, 0), (-2, 0), (-3, 1),$ $(1, 2), (2, -2)$.

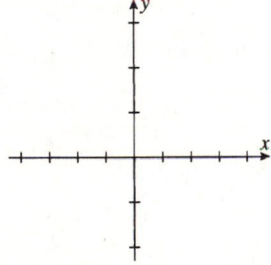

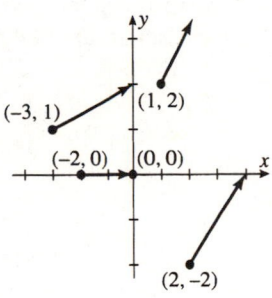

3) Describe and sketch the vector field
$\mathbf{F}(x, y, z) = 2\mathbf{i} + \mathbf{j} + \mathbf{k}$.

To every point in R^3 we associate the (constant) vector $\langle 2, 1, 1 \rangle$. A sketch with the points $(0, 0, 0)$, $(3, 2, 5)$, and $(3, 3, 3)$ is given below.

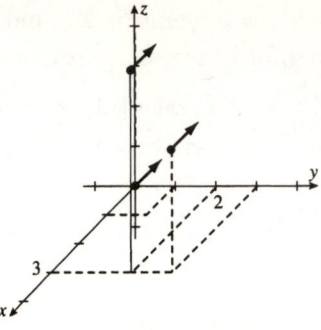

4) Find the gradient vector field of

a) $f(x, y) = x^2 + 3xy$

$\nabla f = (2x + 3y)\mathbf{i} + 3x\mathbf{j}$.

b) $f(x, y, z) = x^2 y + xz^2 + 2yz$

$$\nabla f = (2xy + z^2)\mathbf{i} + (x^2 + 2z)\mathbf{j} \\ + (2xz + 2y)\mathbf{k}$$

5) If **F** is a conservative vector field, then _____ for some scalar function f.

$\nabla f = \mathbf{F}$.

Section 17.2   Line Integrals

This section generalizes the idea of integrating a single-variable real function over an interval $[a, b]$ in two ways: (1) to an integral of a function of two or three variables over a curve in two- or three-dimensional space, and (2) to the integral of a vector field over a curve. Applications of these line integrals include calculating work done moving an object along a curve and finding the mass and center of mass of a wire whose density varies along its length.

Concepts to Master

A. Line integral in two and three dimensions; Piecewise smooth curve; Orientation of a curve

B. Mass and center of mass of a wire

C. Line integral of a vector field; Work

Summary and Focus Questions

Page
1098
(ET Page
1062)

A. Let f be a function of two variables and C be a smooth curve given by $\mathbf{r}(t) = x(t)\mathbf{i} + y(t)\mathbf{j}$, for $t \in [a, b]$. Partitioning $[a, b]$ into n subintervals divides the curve C into subarcs of length Δs_i. By choosing t_i in each subinterval, we obtain sample points (x_i^*, y_i^*) on each subarc. The corresponding Riemann sum is

$$\sum_{i=1}^{n} f(x_i^*, y_i^*)\Delta s_i.$$

The *line integral of f along C* is

$$\int_C f(x, y)ds = \lim_{n \to \infty} \sum_{i=1}^{n} f(x_i^*, y_i^*)\Delta s_i.$$

If f is continuous, $\displaystyle\int_C f(x, y)ds = \int_a^b f(x(t), y(t))\sqrt{\left(\frac{dx}{dt}\right)^2 + \left(\frac{dy}{dt}\right)^2}\, dt.$

Line integrals in 3-dimensional space are defined and computed similarly.

$$\int_C f(x, y, z)ds = \int_a^b f(x(t), y(t), z(t))\sqrt{\left(\frac{dx}{dt}\right)^2 + \left(\frac{dy}{dt}\right)^2 + \left(\frac{dz}{dt}\right)^2}\, dt.$$

A curve C is *piecewise smooth* if C is a union of a finite number of smooth subarcs.

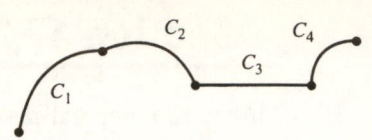

If C is piecewise smooth and composed of smooth curves C_1 and C_2 then

$$\int_C f\, ds = \int_{C_1} f\, ds + \int_{C_2} f\, ds.$$

The line integral of f along C *with respect to x* is

$$\int_C f(x, y)\, dx = \int_a^b f(x(t), y(t))x'(t)dt$$

and *with respect to y* is

$$\int_C f(x, y)\, dy = \int_a^b f(x(t), y(t))y'(t)dt.$$

A parametrization $x = x(t), y = y(t), t \in [a, b]$ of a curve C determines an *orientation* or direction along the curve as t increases from a to b. The curve $-C$ is the same set of points as C but with the reverse orientation.

1) Evaluate $\int_C y\, ds$, where C is given by $x(t) = t, y(t) = t^3, 0 \le t \le \sqrt[4]{7}$.

$x'(t) = 1$ and $y'(t) = 3t^2$, so

$$\int_C y\, ds = \int_0^{\sqrt[4]{7}} t^3 \sqrt{1^2 + (3t^2)^2}\, dt$$

$$= \int_0^{\sqrt[4]{7}} t^3 \sqrt{1 + 9t^4}\, dt$$

$$= \frac{1}{54}(1 + 9t^4)^{3/2}\Big]_{t=0}^{\sqrt[4]{7}} = \frac{511}{54}.$$

2) Set up definite integrals for $\int_C (x - y)dx$ and $\int_C (x - y)dy$ where C is the curve given by $x = 1 + e^t, y = 1 - e^{-t}, 0 \le t \le 1$.

$x'(t) = e^t, y'(t) = e^{-t}.$

$$\int_C (x - y)dx = \int_0^1 (1 + e^t - (1 - e^{-t}))e^t\, dt$$

$$= \int_0^1 (e^{2t} + 1)dt.$$

$$\int_C (x - y)dy = \int_0^1 (1 + e^t - (1 - e^{-t}))e^{-t}\, dt$$

$$= \int_0^1 (1 + e^{-2t})dt.$$

3) Write a definite integral for $\int_C xyz\, ds$ where C is the helix $x = t$, $y = \sin t$, $z = \cos t$, $0 \le t \le \pi$.

$$\int_0^\pi (t \sin t \cos t)\sqrt{1^2 + \cos^2 t + \sin^2 t}\, dt$$

$$= \int_0^\pi \sqrt{2}\, t \sin t \cos t\, dt = \int_0^\pi \frac{t \sin 2t}{\sqrt{2}}\, dt.$$

Page 1100 (ET Page 1064)

B. If $\rho(x, y)$ is the density of the material in a thin wire shaped like a curve C then the *mass* of the wire is

$$m = \int_C \rho(x, y)\, ds.$$

The *center of mass* of the wire is located at the point $(\bar{x}, \bar{y})$ where

$$\bar{x} = \frac{\int_C x\, \rho(x, y)\, ds}{m} \quad \text{and} \quad \bar{y} = \frac{\int_C y\, \rho(x, y)\, ds}{m}.$$

4) Find the center of mass of a piece of wire along the line $y = 2x$ $(0 \le x \le 1)$ whose density at (x, y) is $1 + xy$.

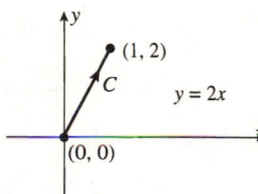

The curve C may be represented by $x = t$, $y = 2t$, $0 \le t \le 1$.

The mass is $m = \int_C (1 + xy)\, ds$

$$= \int_0^1 (1 + t(2t))\sqrt{1^2 + 2^2}\, dt$$

$$= \int_0^1 \sqrt{5}(1 + 2t^2)\, dt = \sqrt{5}\left(t + \tfrac{2}{3}t^3\right)\Big|_0^1$$

$$= \sqrt{5}\left(\left(1 + \tfrac{2}{3}\right) - 0\right) = \frac{5\sqrt{5}}{3}.$$

$$\int_C x\, \rho(x, y)\, ds = \int_0^1 t(1 + 2t^2)\sqrt{5}\, dt$$

$$= \int_0^1 \sqrt{5}(t + 2t^3)\, dt = 5\left(\tfrac{1}{2}t^2 + \tfrac{1}{2}t^4\right)\Big|_0^1$$

$$= 5\left(\left(\tfrac{1}{2} + \tfrac{1}{2}\right) - 0\right) = \sqrt{5}.$$

Thus $\bar{x} = \dfrac{\sqrt{5}}{\frac{5\sqrt{5}}{3}} = \dfrac{3}{5}.$

$$\int_C y \, \rho(x, y) ds = \int_0^1 (2t)(1 + 2t^2)\sqrt{5} \, dt$$
$$= \int_0^1 \sqrt{5}(2t + 4t^3) dt = \sqrt{5}(t^2 + t^4)\Big]_0^1$$
$$= \sqrt{5}[(1 + 1) - 0] = 2\sqrt{5}.$$
$$\text{Thus } \bar{y} = \frac{2\sqrt{5}}{\frac{5\sqrt{5}}{3}} = \frac{6}{5}.$$

Page 1103 (ET Page 1067)

C. If **F** is a continuous vector field on a smooth curve C given by $\mathbf{r}(t)$, $t \in [a, b]$, the *line integral of* **F** *along* C is

$$\int_C \mathbf{F} \cdot d\mathbf{r} = \int_a^b \mathbf{F}(\mathbf{r}(t)) \cdot \mathbf{r}'(t) dt.$$

This may also be written as $\int_C \mathbf{F} \cdot \mathbf{T} \, ds$ where **T** is the unit tangent vector function C, and **F** $\cdot$ **T** is the dot product.

A line integral of a vector field $\mathbf{F} = P\mathbf{i} + Q\mathbf{j} + R\mathbf{k}$ may be written in terms of line integrals of its components:

$$\int_C \mathbf{F} \cdot d\mathbf{r} = \int_C P \, dx + Q \, dy + R \, dz.$$

$\left(\int_C P \, dx + Q \, dy + R \, dz \text{ is shorthand for } \int_C P \, dx + \int_C Q \, dy + \int_C R \, dz. \right)$

If **F** is interpreted as a force field, then $\int_C \mathbf{F} \cdot d\mathbf{r}$ is the *work* done by the force **F** moving a particle along the curve C.

5) Evaluate $\int_C \mathbf{F} \cdot d\mathbf{r}$ where C is the line segment from $(0, 0, 0)$ to $(1, 3, 2)$ and
$$\mathbf{F}(x, y, z) = (y + z)\mathbf{i} + (2x + z)\mathbf{j} + (x + 3y)\mathbf{k}.$$

C may be parametrized by $x = t$, $y = 3t$, $z = 2t$ for $t \in [0, 1]$.

We have $dx = dt$, $dy = 3dt$, $dz = 2dt$.

$$\int_C \mathbf{F} \cdot d\mathbf{r} = \int_C (y + z)dx + (2x + z)dy$$
$$+ (x + 3y)dz$$
$$= \int_0^1 (3t + 2t)dt + (2t + 2t)3dt$$
$$+ (t + 9t)2dt$$
$$\int_0^1 37t \, dt = \frac{37}{2}.$$

6) In the force field $\mathbf{F}(x, y) = x\mathbf{i} + (x + y)\mathbf{j}$, find the amount of work done moving a particle along the parabola $y = x^2$ from $(1, 1)$ to $(2, 4)$.

The parabolic curve C may be described by $x = t, y = t^2, t \in [1, 2]$.
$x'(t) = 1, y'(t) = 2t$.
$dx = dt, dy = 2t\, dt$.

The work done is

$$\int_C \mathbf{F} \cdot d\mathbf{r} = \int_C x\, dx + (x + y)dy$$

$$= \int_1^2 t(1)dt + (t + t^2)2t\, dt$$

$$= \int_1^2 (t + 2t^2 + 2t^3)\, dt = \frac{41}{3}.$$

Section 17.3   The Fundamental Theorem for Line Integrals

The Fundamental Theorem of Calculus for functions of one variable tells us that
$\int_a^b f'(x)dx = f(b) - f(a)$. If you think of ∇f as a higher dimensional analog of
$f'(x)$, then this section uses a similar result to evaluate line integrals of the form
$\int_C \nabla f(x, y)ds$. Certain line integrals (called independent of path) will always be
of this form.

Concepts to Master

A. Line integrals of gradients

B. Closed path; Simple path; Connected; Simply-connected; Independence of path; Test for path independence

Summary and Focus Questions

Page 1110 (ET Page 1074)

A. If C is a piecewise smooth curve given by $\mathbf{r}(t)$, $a \le t \le b$ and f is a differentiable function with ∇f continuous on C, then
$$\int_C \nabla f \cdot d\mathbf{r} = f(\mathbf{r}(b)) - f(\mathbf{r}(a)).$$

1) Let $f(x, y) = x^2 + xy + 2y^2$ and C be any smooth curve from $(0, 1)$ to $(3, 0)$. Find
$$\int_C \nabla f \cdot d\mathbf{r}.$$

$$\int_C \nabla f \cdot d\mathbf{r} = f(3, 0) - f(0, 1) = 9 - 2 = 7.$$

2) If C is a smooth curve given by $\mathbf{r}(t)$ with $\mathbf{r}(a) = \mathbf{r}(b)$ and f is differentiable and ∇f is continuous on C, then $\int_C \nabla f \cdot d\mathbf{r} = $ _____.

0. Since $\mathbf{r}(a) = \mathbf{r}(b)$,
$f(\mathbf{r}(a)) = f(\mathbf{r}(b))$

B. A curve C given by $\mathbf{r}(t)$, $a \le t \le b$ is *closed* means $\mathbf{r}(a) = \mathbf{r}(b)$.

A curve C is *simple* means C does not intersect itself except, perhaps, at its endpoints.

Simple, not closed   Closed, not simple

Closed and simple

A domain D is *open* if every point is an interior point. D is *connected* if any two points of D can be connected by a path entirely in D. D is *simply-connected* means every simple closed curve in D encloses only points of D.

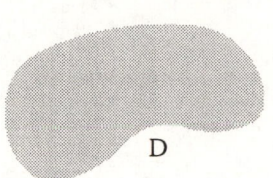

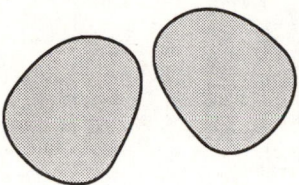

Open
(no boundary points in D).

Not Connected

Connected, not
simply-connected

For a continuous vector function $\mathbf{F}$ with domain D,

$\int_C \nabla f \, d\mathbf{r}$ is *independent of path* means its value depends only upon the endpoints A and B and not on any particular path from A to B.

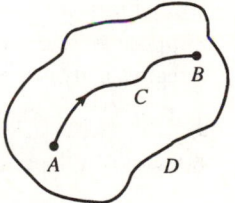

The results in part **A** of this section say that

$\int_C \nabla f \cdot d\mathbf{r}$ is independent of path. Conversely, under certain conditions, every independent of path line integral is a gradient line integral.

Let $\mathbf{F}$ be continuous on an open connected region D. If $\int_C \mathbf{F} \, d\mathbf{r}$ is independent of path, then $\mathbf{F}$ is conservative (i.e., $\mathbf{F} = \nabla f$, for some differentiable f).

For an open simply-connected region D, if $\dfrac{\partial P}{\partial y}$ and $\dfrac{\partial Q}{\partial x}$ are continuous on D, then $\mathbf{F} = P\mathbf{i} + Q\mathbf{j}$ is the gradient of some function f if and only if $\dfrac{\partial P}{\partial y} = \dfrac{\partial Q}{\partial x}$. Thus $\dfrac{\partial P}{\partial y} = \dfrac{\partial Q}{\partial x}$ is a test of whether $\mathbf{F}$ is conservative (and thus path independence).

If $\mathbf{F} = P\mathbf{i} + Q\mathbf{j} = \nabla f$, to reconstruct f, integrate $\int P \, dx$, differentiate the result with respect to y, and then compare that result with Q. Or, you can compare P with $\dfrac{\partial}{\partial x}(\int Q \, dy)$.

3)

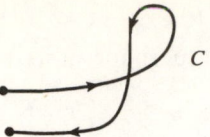

C

True, False:

a) C is simple

False.

b) C is closed

False.

4) True, False:
The region below is simply connected.

True.

5) If **F** is conservative and C is a smooth closed curve then $\displaystyle\int_C \mathbf{F} \cdot d\mathbf{r} = $ _____.

0. (For some f, $\mathbf{F} = \nabla f$, so $\displaystyle\int_C \mathbf{F} \cdot d\mathbf{r} = \int_C \nabla f \cdot d\mathbf{r} = 0$ since the curve is closed.)

6) Suppose C_1 and C_2 are two distinct paths in a region D from point A to point B. If $\displaystyle\int_{C_1} \mathbf{F}\, d\mathbf{r} = \int_{C_2} \mathbf{F}\, d\mathbf{r}$, can we conclude the line integral is independent of path?

No, independence requires every line integral along all paths in D from A to B to have the same value.

7) Is $\mathbf{F}(x, y) = (2x + y)\mathbf{i} + (x + 8y)\mathbf{j}$ the gradient of some function?

Yes. For $P(x, y) = 2x + y$ and $Q(x, y) = x + 8y$, $\dfrac{\partial P}{\partial y} = 1 = \dfrac{\partial Q}{\partial x}$, so $\mathbf{F} = \nabla f$ for some f.

8) Find a function f whose gradient is
$\nabla f = (y^2 + 1)\mathbf{i} + (2xy + 3y^2)\mathbf{j}$.

Let $P(x, y) = y^2 + 1$ and
$Q(x, y) = 2xy + 3y^3$.
$f(x, y) = \int P\, dx = \int (y^2 + 1)\, dx$
$= xy^2 + x + h(y)$,
where h is a function of y.
$\dfrac{\partial f}{\partial y} = 2xy + h'(y)$. Comparing to Q,
$h'(y) = 3y^2$ so $h(y) = y^3 + K$.
Thus $f(x, y) = xy^2 + x + y^3 + K$.
For $K = 0$, $f(x, y) = xy^2 + x + y^3$.

9) Let C be the parabola $x = y^2$ from $(1, 1)$ to $(4, 2)$. Evaluate
$$\int_C (2x + 3y^2)dx + (6xy + 10)dy.$$

Let $P(x, y) = 2x + 3y^2$ and
$Q(x, y) = 6xy + 10$.
Since $\dfrac{\partial P}{\partial y} = 6y = \dfrac{\partial Q}{\partial x}$, $P\mathbf{i} + Q\mathbf{j}$ is a gradient
of some function $f(x, y)$.
$f(x, y) = \int (2x + 3y^2)dx$
$\qquad = x^2 + 3xy^2 + h(y)$.
$\dfrac{\partial f}{\partial y} = 6xy + h'(y)$.
Comparing with $Q(x, y)$, $h'(y) = 10$, so
$h(y) = 10y + c$. We choose $c = 0$. Thus
$f(x, y) = x^2 + 3xy^2 + 10y$ and
$$\int_C P\, dx + Q\, dy = f(4, 2) - f(1, 1)$$
$$= 84 - 14 = 70.$$
(Note that it did not matter that C was a parabola.)

10) Evaluate $\int_C dx + z\, dy + y\, dz$ where C is any path from $(0, 1, 1)$ to $(3, 1, 6)$.

Let $P(x, y, z) = 1$, $Q(x, y, z) = z$, and
$R(x, y, z) = y$.
$f(x, y, z) = \int P\, dx = x + C(y, z)$.
$f_y = C_y(y, z) = Q = z$.
So $C(y, z) = yz + D(z)$.
Thus $f(x, y, z) = x + yz + D(z)$.
$f_z = y + D'(z) = R = y$. Thus
$D'(z) = 0$, so $D(z) = K$. We use $K = 0$.
Therefore $f(x, y, z) = x + yz$ and
$$\int_C dx + z\, dy + y\, dz = (x + yz)\Big|_{(0, 1, 1)}^{(3, 1, 6)}$$
$$= (3 + 6) - (0 + 1) = 8.$$

Section 17.4 Green's Theorem

This section includes a handy relationship between a double integral over a plane region D and a line integral around the boundary curve for D. It also provides an alternate method to calculate the area of a region D.

Concepts to Master

A. Positive orientation of a simple closed curve; Green's Theorem

B. Evaluate areas with line integrals

Summary and Focus Questions

Page 1119 (ET Page 1083)

A. Let C be a simple closed curve in the plane given by $\mathbf{r}(t)$, for $a \le t \le b$. We say C has a *positive orientation* if the traversal of C from $\mathbf{r}(a)$ to $\mathbf{r}(b)$ is counterclockwise. In this case the line integral for a vector $\mathbf{F}$ about C is denoted $\oint_C \mathbf{F} \cdot d\mathbf{r}$.

Green's Theorem: Let C be a piecewise smooth, simple closed curve with positive orientation about a region D. If $P(x, y)$ and $Q(x, y)$ have continuous partials in an open region containing D, then

$$\oint_C P \, dx + Q \, dy = \iint_D \left(\frac{\partial Q}{\partial x} - \frac{\partial P}{\partial y} \right) dA.$$

1) Place arrows on the curve C near the point x to indicate a positive orientation to the curve.

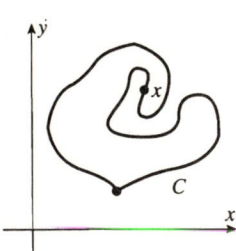

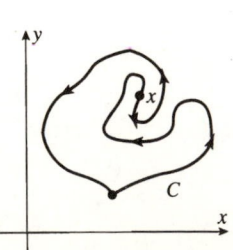

2) Evaluate using Green's Theorem:

a) $\oint_C xy\,dx + y^2dy$, where C is the triangle formed by $(0, 0)$, $(1, 1)$, and $(0, 1)$.

Let $P(x, y) = xy$, $Q(x, y) = y^2$.

Then $\dfrac{\partial P}{\partial y} = x$ and $\dfrac{\partial Q}{\partial x} = 0$.

C is pictured below and encloses the region $D: 0 \le x \le 1, x \le y \le 1$.

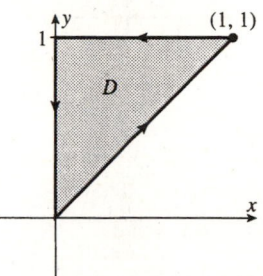

The line integral becomes

$$\iint_D (0 - x)dx\,dy = \int_0^1\int_x^1 -x\,dy\,dx$$

$$= \int_0^1 (x^2 - x)dx = -\frac{1}{6}.$$

b) $\oint_C (6xy^2)dx + (6x^2y)dy$, where C is the triangle in part a).

For $P(x, y) = 6xy^2$ and $Q(x, y) = 6x^2y$,

$$\frac{\partial P}{\partial y} = 12xy = \frac{\partial Q}{\partial x}.$$

Thus $\dfrac{\partial Q}{\partial x} - \dfrac{\partial P}{\partial y} = 0$,

so $\oint_C P\,dx + Q\,dy = \iint_D 0\,dA = 0.$

Page
1122
(ET Page
1086)

B. By selecting P and Q such that $\dfrac{\partial Q}{\partial x} - \dfrac{\partial P}{\partial y} = 1$ (there are many choices), three equivalent line integrals may be used to find the area of D:

Area of $D = \oint_C x\,dy = -\oint_C y\,dx = \frac{1}{2}\oint x\,dy - y\,dx.$

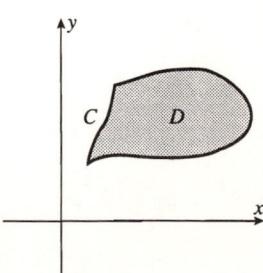

3) Find a line integral expression for the region D:

a)

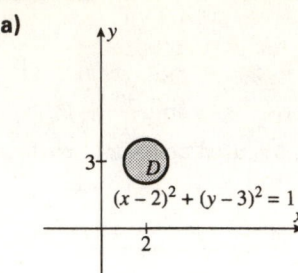

$$(x-2)^2 + (y-3)^2 = 1$$

We know the answer will be π since D is a circle of radius 1. The boundary C may be described as $x = 2 + \cos\theta$, $y = 3 + \sin\theta$, $0 \le \theta \le 2\pi$. The area of D is

$$\frac{1}{2}\int x\,dy - y\,dx$$

$$= \frac{1}{2}\int_0^{2\pi} (2 + \cos\theta)\cos\theta$$
$$-(3 + \sin\theta)(-\sin\theta)d\theta$$

$$= \frac{1}{2}\int_0^{2\pi} (2\cos\theta + 3\sin\theta + 1)d\theta$$

$$= \frac{1}{2}(2\sin\theta - 3\cos\theta + \theta)\Big|_0^{2\pi} = \pi.$$

Section 17.5   Curl and Divergence

This section describes two operations in three dimensions that resemble differentiation and may be used to describe fluid flow. The curl at a point produces a vector that measures the rotation of fluid about a point in the flow. The div at a point measures the rate of change of fluid flowing from the point (how fast is the fluid *div*erging from the point).

Concepts to Master

A. Del operator; Curl of a vector field

B. Divergence of a vector field

C. Vector forms of Green's Theorem

Summary and Focus Questions

Page 1126 (ET Page 1090)

A. The operations in this section result in either a vector field (curl **F**) or a scalar field (div **F**).

The ∇ *(del)* operator is the expression $\nabla = \mathbf{i}\dfrac{\partial}{\partial x} + \mathbf{j}\dfrac{\partial}{\partial y} + \mathbf{k}\dfrac{\partial}{\partial z}$.

The *curl of* $\mathbf{F} = P\mathbf{i} + Q\mathbf{j} + R\mathbf{k}$ is

$$\text{curl } \mathbf{F} = \nabla \times \mathbf{F} = \begin{vmatrix} \mathbf{i} & \mathbf{j} & \mathbf{k} \\ \dfrac{\partial}{\partial x} & \dfrac{\partial}{\partial y} & \dfrac{\partial}{\partial z} \\ P & Q & R \end{vmatrix}$$

$$= \left(\frac{\partial R}{\partial y} - \frac{\partial Q}{\partial z}\right)\mathbf{i} + \left(\frac{\partial P}{\partial z} - \frac{\partial R}{\partial x}\right)\mathbf{j} + \left(\frac{\partial Q}{\partial x} - \frac{\partial P}{\partial y}\right)\mathbf{k}.$$

The curl **F** measures a rate of change of circulation – if **F** is the velocity of a fluid flow, then curl $\mathbf{F}(x, y, z)$ measures the circulation per unit area orthogonal to $\mathbf{F}(x, y, z)$ at (x, y, z). If curl $\mathbf{F} = 0$ (irrotation) at a point, there is no circulation about the point as fluid flows through the point with force **F**.

If $f(x, y, z)$ has continuous second partials, curl $(\nabla f) = \mathbf{0}$, the zero vector. Conversely, if $\mathbf{F}(x, y, z)$ has components with continuous partial derivatives and curl $\mathbf{F} = \mathbf{0}$ then **F** is conservative.

1) Find the curl of
$\mathbf{F}(x, y, z) = xy^2\mathbf{i} + x^2z\mathbf{j} + yz^2\mathbf{k}$.

Let $P(x, y, z) = xy^2$, $Q(x, y, z) = x^2z$, and
$R(x, y, z) = yz^2$.

$\dfrac{\partial P}{\partial y} = 2xy, \dfrac{\partial P}{\partial z} = 0, \dfrac{\partial Q}{\partial x} = 2xz, \dfrac{\partial Q}{\partial z} = x^2,$

$\dfrac{\partial R}{\partial x} = 0, \dfrac{\partial R}{\partial y} = z^2.$

$\operatorname{curl} \mathbf{F} = (z^2 - x^2)\mathbf{i} + (0 - 0)\mathbf{j}$
$\qquad\qquad + (2xz - 2xy)\mathbf{k}$

$\qquad = (z^2 - x^2)\mathbf{i} - (2xz - 2xy)\mathbf{k}.$

2) Use the curl $\mathbf{F}$ to determine whether
$\mathbf{F}(x, y, z) = xy\mathbf{i} + z^2\mathbf{j} + xz\mathbf{k}$ is conservative.

$\dfrac{\partial P}{\partial y} = x, \dfrac{\partial P}{\partial z} = 0, \dfrac{\partial Q}{\partial x} = 0, \dfrac{\partial Q}{\partial z} = 2z,$

$\dfrac{\partial R}{\partial x} = z, \dfrac{\partial R}{\partial y} = 0.$

$\operatorname{curl} \mathbf{F} = (0 - 2z)\mathbf{i} + (0 - z)\mathbf{j} + (0 - x)\mathbf{k}$
$\qquad = -2z\mathbf{i} - z\mathbf{j} - x\mathbf{k}$, not the zero vector.

Therefore, $\mathbf{F}$ is not conservative.

3) Show that $\mathbf{F}(x, y, z)$
$= (2x + y)\mathbf{i} + (x + 2yz^2)\mathbf{j} + 2y^2z\mathbf{k}$ is
conservative and find any f such that
$\nabla f = \mathbf{F}$.

$\dfrac{\partial P}{\partial y} = 1, \dfrac{\partial P}{\partial z} = 0, \dfrac{\partial Q}{\partial x} = 1, \dfrac{\partial Q}{\partial z} = 4yz,$

$\dfrac{\partial R}{\partial x} = 0, \dfrac{\partial R}{\partial y} = 4yz.$

$\operatorname{curl} \mathbf{F} = (4yz - 4yz)\mathbf{i} + (0 - 0)\mathbf{j}$
$\qquad\qquad + (1 - 1)\mathbf{k} = \mathbf{0}$

so $\mathbf{F}$ is conservative.

$f_x = 2x + y, f_y = x + 2yz^2, f_z = 2y^2z.$

From $f_x = 2x + y,$
$f(x, y, z) = x^2 + xy + C(y, z).$
$f_y = x + C_y(y, z) = x + 2yz^2.$

Thus $C_y(y, z) = 2yz^2,$
so $C(y, z) = y^2z^2 + K(z).$

Hence $f(x, y, z) = x^2 + xy + y^2z^2 + K(z).$

Now $f_z = 2y^2z + K_z(z) = 2y^2z$, so $K_z(z) = 0$. Choose $K(z) = 0.$

Therefore $f(x, y, z) = x^2 + xy + y^2z^2.$

Page 1129 (ET Page 1093)

B. For $\mathbf{F} = P\mathbf{i} + Q\mathbf{j} + R\mathbf{k}$, *the divergence of $\mathbf{F}$ is*

$$\text{div } \mathbf{F} = \nabla \cdot \mathbf{F} = \frac{\partial P}{\partial x} + \frac{\partial Q}{\partial y} + \frac{\partial R}{\partial z}.$$

If $\mathbf{F}$ is the velocity of a fluid flow, then div $\mathbf{F}(x, y, z)$ may be interpreted as the (instantaneous) rate of change of the mass of the fluid per unit volume. Thus if div $\mathbf{F}(x, y, z) > 0$ there is a net flow out of the point (x, y, z). (The fluid is diverging.)

For $\mathbf{F} = P\mathbf{i} + Q\mathbf{j} + R\mathbf{k}$, with continuous second partial derivatives, div curl $\mathbf{F} = 0$.

4) Find div $\mathbf{F}$ for $\mathbf{F}(x, y, z) = (x^2 + y)\mathbf{i} + (y^2 + z)\mathbf{j} + xyz\mathbf{j}$.

$$\frac{\partial P}{\partial z} = 2x, \frac{\partial Q}{\partial y} = 2y, \frac{\partial R}{\partial z} = xy.$$
$$\text{div } \mathbf{F} = 2x + 2y + xy.$$

5) True, False: curl div $\mathbf{F} = 0$.

False. For a vector field $\mathbf{F}$, div $\mathbf{F}$ is a scalar. Curl can only be computed for a vector field, so curl div $\mathbf{F}$ makes no sense.

6) For a fluid with velocity $\mathbf{F} = 2\mathbf{i} + y^2\mathbf{j} + (x + z)\mathbf{k}$, is the fluid diverging at $(1, 2, 1)$?

Yes. div $\mathbf{F} = \dfrac{\partial P}{\partial x} + \dfrac{\partial Q}{\partial y} + \dfrac{\partial R}{\partial z} = 0 + 2y + 1$
$= 2y + 1$.

At $(1, 2, 1)$, div $\mathbf{F} = 2(2) + 1 = 5$.

Since div $\mathbf{F} > 0$ the fluid is diverging.

Page 1131 (ET Page 1095)

C. Using the concepts of div and curl, Green's Theorem may be expressed in vector forms:

(1) $\displaystyle\oint_C \mathbf{F} \cdot d\mathbf{r} = \iint_D (\text{curl } \mathbf{F}) \cdot \mathbf{k}\, dA$

(2) $\displaystyle\oint_C \mathbf{F} \cdot \mathbf{n}\, d\mathbf{r} = \iint \text{div } \mathbf{F}(x, y)\, dA$

where $\mathbf{n}$ is the outer normal to the tangent vector $\mathbf{T}$ to C,

$$\mathbf{n}(t) = \frac{y'(t)}{|\mathbf{r}'(t)|}\mathbf{i} - \frac{x'(t)}{|\mathbf{r}'(t)|}\mathbf{j}.$$

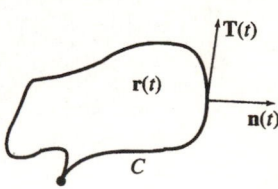

7) Write a double integral for $\oint_C \mathbf{F} \cdot d\mathbf{r}$ where C is the curve $\dfrac{x^2}{4} + \dfrac{y^2}{9} = 1$ and $\mathbf{F}(x, y, z) = e^{xy}\mathbf{i} - e^{xy}\mathbf{j} + \mathbf{k}$.

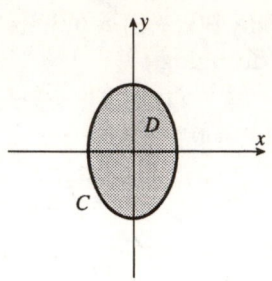

$$P(x, y, z) = e^{xy}$$
$$Q(x, y, z) = -e^{xy}$$
$$R(x, y, z) = 1$$

(1) $\operatorname{curl} \mathbf{F} = \begin{vmatrix} \mathbf{i} & \mathbf{j} & \mathbf{k} \\ \dfrac{\partial}{\partial x} & \dfrac{\partial}{\partial y} & \dfrac{\partial}{\partial z} \\ e^{xy} & -e^{xy} & 1 \end{vmatrix}$

$$= (0 - 0)\mathbf{i} + (0 - 0)\mathbf{j} + (-ye^{xy} - xe^{xy})\mathbf{k}$$
$$= (-ye^{xy} - xe^{xy})\mathbf{k}.$$

$\operatorname{curl} \mathbf{F} \cdot \mathbf{k} = ((-ye^{xy} - xe^{xy})\mathbf{k}) \cdot \mathbf{k}$
$$= -ye^{xy} - xe^{xy} \text{ (since } \mathbf{k} \cdot \mathbf{k} = 1).$$

Thus the line integral is

$$\iint_D (-ye^{xy} - xe^{xy})dA.$$

(2) $\operatorname{div} \mathbf{F} = \dfrac{\partial P}{\partial x} + \dfrac{\partial Q}{\partial y} + \dfrac{\partial R}{\partial z}$
$$= ye^{xy} - xe^{xy}$$

The line integral may also be written as

$$\iint_D (ye^{xy} - xe^{xy})dA.$$

Section 17.6 Parametric Surfaces and Their Areas

Surfaces in space are two-dimensional objects. This section shows how to parameterize a surface in a manner similar to the way curves are parameterized—we need two parametric variables instead of one. You will also learn how to calculate tangent planes and areas for parametric surfaces.

Concepts to Master

A. Parameterization of a surface; Grid curve

B. Tangent plane to a parametric surface

C. Area of a parametric surface

Summary and Focus Questions

Page 1134 (ET Page 1098)

A. Let $\mathbf{r}(u, v) = x(u, v)\mathbf{i} + y(u, v)\mathbf{j} + z(u, v)\mathbf{k}$ be a vector function with domain D, a subset of the uv-plane. The set of all points $\mathbf{r}(u, v)$, where (u, v) varies throughout D, is a *parametric surface*, S, in three dimensions. The equations

$$x = x(u, v) \qquad y = y(u, v) \qquad z = z(u, v)$$

parameterize the surface S.

Example: Let D be the triangular region below in the uv-plane and let

$$\mathbf{r}(u, v) = (3u)\mathbf{i} + (2v)\mathbf{j} + (6 - 6u - 6v)\mathbf{k}.$$

The resulting parameterized surface, S, is that portion of the plane $z = 6 - 2x - 3y$ in the first octant of xyz-space.

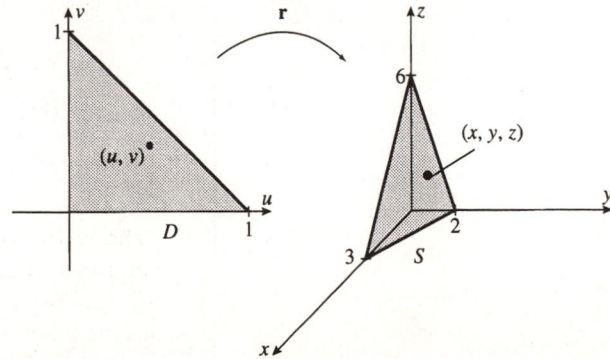

By keeping u constant, $u = u_0$, the function $\mathbf{r}(u_0, v)$ becomes a vector function of the one parameter v and determines a *grid curve* on the surface S. Similarly, by keeping $v = v_0$, we get another grid curve determined by $\mathbf{r}(u, v_0)$.

Grid curves help visualize the parameterization. For the above example, $u = \frac{1}{3}$ and $u = \frac{2}{3}$ produce two grid lines on the surface S.

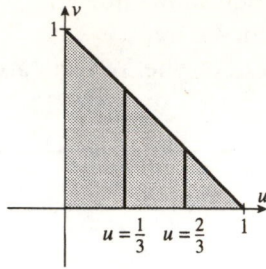 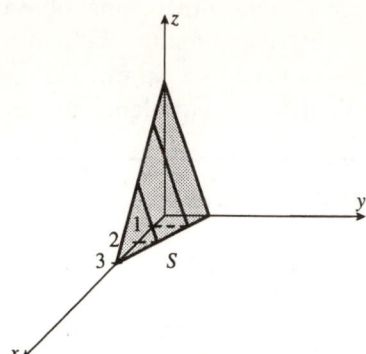

1) Parameterize the cone $z^2 = x^2 + y^2$, $0 \le z \le 1$, using polar coordinates.

Using polar coordinates (r, θ), let D be the unit circle $0 \le \theta \le 2\pi, 0 \le r \le 1$. Then $x = r \cos \theta, y = r \sin \theta, z = r, 0 \le \theta \le 2\pi,$ $0 \le r \le 1$ parameterizes the cone.

2) Parameterize the surface of revolution obtained by revolving $y = x^2, 0 \le x \le 2$ about the x-axis.

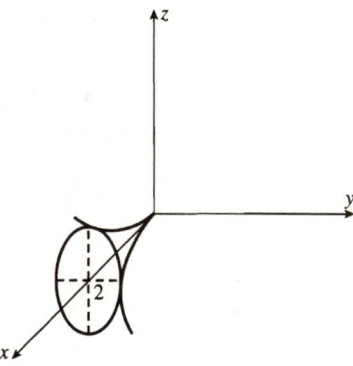

The surface is about the x-axis so we use x as one parameter: $x = x$. For any x, the surface is a circle with radius x^2, so we use θ as the other parameter: $y = x^2 \cos \theta$, $z = x^2 \sin \theta$. Our parametrization is $x = x$, $y = x^2 \cos \theta, z = x^2 \sin \theta$ with $0 \le x \le 2$, $0 \le \theta \le 2\pi$.

3) Describe the grid curves for
$$x = u$$
$$y = u^2 + v^2$$
$$z = v$$

For $u = u_0$, $y = u_0^2 + v^2$ is a parabola in a plane parallel to the xy-plane.

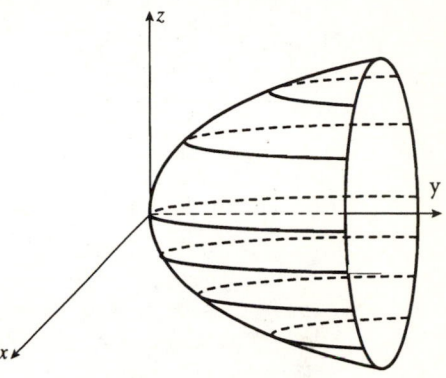

For $v = v_0$, $y = u^2 + v_0^2$ is also a parabola, but in a plane parallel to the yz-plane.

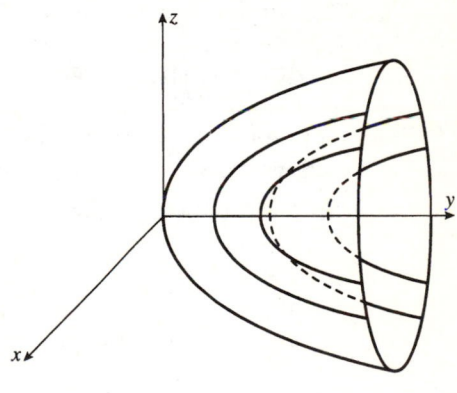

B. The tangent plane at (x_0, y_0, z_0) to a surface S parametrized by $\mathbf{r}(u, v)$ has as a normal vector $\mathbf{r}_u \times \mathbf{r}_v$. The equation of the tangent plane is

$$(\mathbf{r}_u \times \mathbf{r}_v) \cdot \langle x - x_0, y - y_0, z - z_0 \rangle = 0.$$

S is *smooth* if the normal vector is not **0**.

Page
1138
(ET Page
1102)

4) Find the equation of the plane tangent to the surface S at $(u, v) = (1, 1)$ where S is parametrized by

$x = 2u^2v^2$
$y = u + 2v$
$z = 2u + v.$

$\mathbf{r}(u, v) = x\mathbf{i} + y\mathbf{j} + z\mathbf{k}.$

$\mathbf{r}_u = \dfrac{\partial x}{\partial u}\mathbf{i} + \dfrac{\partial y}{\partial u}\mathbf{j} + \dfrac{\partial z}{\partial u}\mathbf{k} = 4uv^2\mathbf{i} + 1\mathbf{j} + 2\mathbf{k}.$

$\mathbf{r}_v = 4u^2v\mathbf{i} + 2\mathbf{j} + 1\mathbf{k}.$

$$\mathbf{r}_u \times \mathbf{r}_v = \begin{vmatrix} \mathbf{i} & \mathbf{j} & \mathbf{k} \\ 4uv^2 & 1 & 2 \\ 4u^2v & 2 & 1 \end{vmatrix}$$

$\qquad = -3\mathbf{i} + (8u^2v - 4uv^2)\mathbf{j}$
$\qquad\qquad + (8uv^2 - 4u^2v)\mathbf{k}.$

At $(u, v) = (1, 1),$
$\mathbf{r}_u \times \mathbf{r}_v = -3\mathbf{i} + 4\mathbf{j} + 4\mathbf{k}$
and $(x, y, z) = (2, 3, 3).$

Thus the tangent plane is
$-3(x - 2) + 4(y - 3) + 4(z - 3) = 0$, or
$3x - 4y - 4z + 18 = 0.$

Page 1140 (ET Page 1104)

C. The surface area of a surface S parametrized by $\mathbf{r}(u, v)$ is

$$A(S) = \iint_D |\mathbf{r}_u \times \mathbf{r}_v|\, dA = \iint_D \sqrt{\left[\dfrac{\partial(x, y)}{\partial(u, v)}\right]^2 + \left[\dfrac{\partial(y, z)}{\partial(u, v)}\right]^2 + \left[\dfrac{\partial(x, z)}{\partial(u, v)}\right]^2}\, dA.$$

5) Set up an iterated integral for the area of each surface S.

a) S is parameterized by
$x = 2uv$
$y = u + v$
$z = u - v$
$0 \le u \le 2, 0 \le v \le 1.$

Let D be the region $0 \le u \le 2, 0, \le v \le 1.$

$\dfrac{\partial(x, y)}{\partial(u, v)} = \dfrac{\partial x}{\partial u}\dfrac{\partial y}{\partial v} - \dfrac{\partial x}{\partial v}\dfrac{\partial y}{\partial u}$

$\qquad = (2v)(1) - (2u)(1) = 2v - 2u.$

$\dfrac{\partial(x, z)}{\partial(u, v)} = (2v)(-1) - (2u)(1) = -2v - 2u.$

$\dfrac{\partial(y, z)}{\partial(u, v)} = (1)(-1) - (1)(1) = -2.$

This area is

$$\int_0^2 \int_0^1 \sqrt{(2v-2u)^2 + (-2v-2u)^2 + (-2)^2}\, dv\, du$$

$$= \int_0^2 \int_0^1 2\sqrt{2u^2 + 2v^2 + 1}\, dv\, du.$$

b) S is the surface
$z = x^2 - y^2$,
$0 \le x \le 1, 0 \le y \le 1$.

Parameterize S with $x = u, y = v$,
$z = u^2 - v^2$.

$$\frac{\partial(x, y)}{\partial(u, v)} = 1(1) - (0)0 = 1$$

$$\frac{\partial(x, z)}{\partial(u, v)} = 1(-2v) - 0(2u) = -2v$$

$$\frac{\partial(y, z)}{\partial(u, v)} = 0(-2v) - 1(2u) = -2u$$

Since $x = u$ and $y = v$ we have $0 \le u \le 1$, $0 \le v \le 1$.

The area is

$$\int_0^1 \int_0^1 \sqrt{1^2 + (-2v^2) + (-2u)^2}\, dv\, du$$

$$= \int_0^1 \int_0^1 \sqrt{1 + 4v^2 + 4u^2}\, dv\, du.$$

Section 17.7   Surface Integrals

In this section we will see that a surface integral is the two-dimensional version of a line integral. Just as a line integral was used to determine the length, mass, etc. of a curve, a surface integral will be used to calculate the area, mass, etc. of a surface.

Concepts to Master

A. Surface integral

B. Orientation of a surface; Surface integral of a vector field

Summary and Focus Questions

Page 1147 (ET Page 1111)

A. Suppose S is a surface parameterized by
$$\mathbf{r}(u, v) = x(u, v)\mathbf{i} + y(u, v)\mathbf{j} + z(u, v)\mathbf{k}$$
with domain D in the uv-plane. The *surface integral* of a function $f(x, y, z)$ whose domain includes S is
$$\iint_S f(x, y, z)dS = \iint_D f(\mathbf{r}(u, v))|\mathbf{r}_u \times \mathbf{r}_v|dA.$$
In the special case where S is parametrized by a function $z = g(x, y)$,
$$\iint_S f(x, y, z)dS = \iint_D f(x, y, g(x, y))\sqrt{(g_x)^2 + (g_y)^2 + 1}\, dA.$$

1) Find an iterated integral for the surface integral of $f(x, y, z) = x$ and S is that portion of the plane $6x + 3y + 2z = 12$ in the first octant.

A parametrization of S is
$$\mathbf{r}(u, v) = u\mathbf{i} + v\mathbf{j} + \left(6 - 3u - \frac{3}{2}v\right)\mathbf{k},$$
for $(u, v) \in D, 0 \le u \le 2, 0 \le v \le 4 - 2u$.
See the figure.

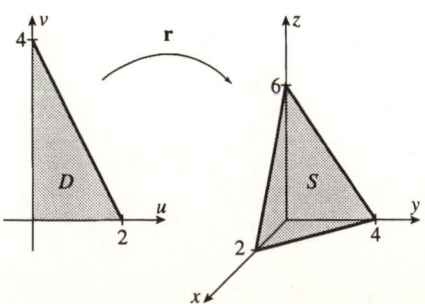

$$r_u \times r_v = \begin{vmatrix} i & j & k \\ 1 & 0 & -\frac{3}{2} \\ 0 & 1 & -\frac{3}{2} \end{vmatrix} = 3i + \frac{3}{2}j + k$$

$$|r_u \times r_v| = \sqrt{3^2 + \left(\frac{3}{2}\right)^2 + 1^2} = \frac{7}{2}.$$

$$\iint_S x\, dS = \iint_D u\left(\frac{7}{2}\right)dA = \int_0^2 \int_0^{4-2u} \frac{7u}{2}\, dv\, du.$$

2) Evaluate $\iint_S y^2 dS$ where S is that part of the cylinder $x^2 + z^2 = 1$ between the planes $y = 0$ and $y = 3 - x$.

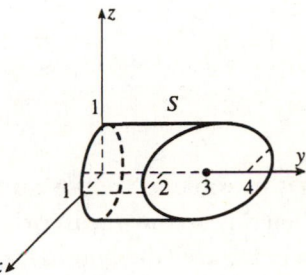

We parameterize S with
$x = \cos u$
$y = v$
$z = \sin u,$
where $0 \le u \le 2\pi$, and $0 \le v \le 3 - \cos u.$

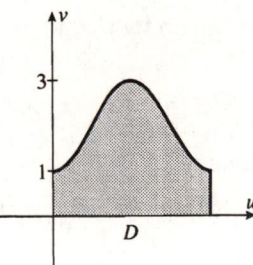

$r_u = \langle -\sin u, 0, \cos u \rangle$ and $r_v = \langle 0, 1, 0 \rangle.$

$$r_u \times r_v = \begin{vmatrix} i & j & k \\ -\sin u & 0 & \cos u \\ 0 & 1 & 0 \end{vmatrix}$$
$$= \cos u\, i - \sin u\, k.$$
$$|r_u \times r_v| = \sqrt{(\cos u)^2 + 0^2 + (-\sin u)^2}$$
$$= 1.$$

Thus $\displaystyle\iint_S y^2\, dS = \int_0^{2\pi}\int_0^{3-\cos u} v^2(1)\, dv\, du$

$\displaystyle = \int_0^{2\pi} \left(\tfrac{1}{3}v^3\right)\Big|_0^{3-\cos u}\ \Big)\, du$

$\displaystyle = \frac{1}{3}\int_0^{2\pi} (3-\cos u)^3\, du$

$\displaystyle = \frac{1}{3}\int_0^{2\pi} (27 - 27\cos u + 9\cos^2 u - \cos^3 u)\, du$

$\displaystyle = \frac{1}{3}\Big(27u - 27\sin u + 9\big[\tfrac{1}{2}u + \tfrac{1}{4}\sin 2u\big]$

$\displaystyle \qquad\qquad - \tfrac{1}{3}[(2+\cos^2 u)\sin u]\big)\Big]_0^{2\pi}$

$\displaystyle = 21\pi.$

B. If it is possible to choose a unit normal vector **n** for each point of a surface S, then S is *oriented* with orientation provided by the given choice of **n**. For a closed surface (the boundary of a solid) a *positive orientation* points outward from the solid.

Page 1151 (ET Page 1115)

If $\mathbf{F} = P\mathbf{i} + Q\mathbf{j} + R\mathbf{k}$ is a continuous vector field whose domain includes an oriented surface S (with unit normal **n**), the *surface integral* or *flux of F* over S defined as

$$\iint_S \mathbf{F}\cdot d\mathbf{S} = \iint_S \mathbf{F}\cdot\mathbf{n}\, dS.$$

When S is given by $\mathbf{r}(u,v) = x(u,v)\mathbf{i} + y(u,v)\mathbf{j} + z(u,v)\mathbf{k}$, for (u,v) in a domain D

$$\iint_S \mathbf{F}\cdot d\mathbf{S} = \iint_D \mathbf{F}\cdot(\mathbf{r}_u \times \mathbf{r}_v)\, dA$$

and in the special case where S is given by $z = g(x,y)$,

$$\iint_S \mathbf{F}\cdot d\mathbf{S} = \iint_D \left(-P\frac{\partial g}{\partial x} - Q\frac{\partial g}{\partial y} + R\right) dA.$$

3) Find an iterated integral for $\iint\limits_{S} \mathbf{F} \cdot d\mathbf{S}$

where S is the top half of the sphere
$x^2 + y^2 + z^2 = 36$ and $\mathbf{F}(x, y, z) = z\mathbf{k}$.

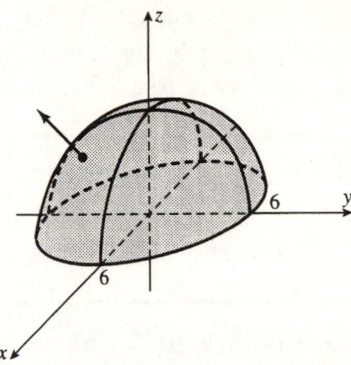

In spherical coordinates, the hemisphere is
$\rho = 6$ with $0 \leq \phi \leq \dfrac{\pi}{2}$.

A parametrization is $x = 6 \sin \phi \cos \theta$,
$y = 6 \sin \phi \sin \theta$, $z = 6 \cos \phi$, $0 \leq \phi \leq \dfrac{\pi}{2}$,
$0 \leq \theta \leq 2\pi$.

$\mathbf{r}_\phi \times \mathbf{r}_\theta$

$$= \begin{vmatrix} \mathbf{i} & \mathbf{j} & \mathbf{k} \\ 6 \cos \phi \cos \theta & 6 \cos \phi \sin \theta & -6 \sin \phi \\ -6 \sin \phi \sin \theta & 6 \sin \phi \cos \theta & 0 \end{vmatrix}$$

$= 36(\sin^2\phi \cos \theta \mathbf{i} + \sin^2\phi \sin \theta \mathbf{j}$
$\quad + \sin\phi \cos \phi \mathbf{k})$.

$\mathbf{F} \cdot (\mathbf{r}_\phi \times \mathbf{r}_\theta) = z(36 \sin \phi \cos \phi)$

$\quad = 6 \cos \phi \, 36 \sin \phi \cos \phi$

$\quad = 216 \sin \phi \cos^2\phi$.

$$\iint\limits_{S} \mathbf{F} \cdot d\mathbf{S} = \int_0^{\pi/2} \int_0^{2\pi} 216 \sin \phi \cos^2\phi \; d\theta \; d\phi.$$

Section 17.8 Stokes' Theorem

This section and the next provide extensions of Green's Theorem. Stokes' Theorem may be thought of as Green's Theorem in three dimensions. Rather than relate a double over a (flat) region in the plane to a line integral around the boundary curve, Stokes' Theorem relates a surface integral for a surface S in three-dimensions to a line-integral around the boundary curve of S.

Concepts to Master

Stokes' Theorem

Summary and Focus Questions

Page 1157 (ET Page 1121)

Stokes' Theorem: Let S be a smooth, simply connected, orientable surface bounded by a simple closed curve C. Let $\mathbf{F} = P\mathbf{i} + Q\mathbf{j} + R\mathbf{k}$, where P, Q, and R are component functions with continuous first partial derivatives. Finally, suppose C has a positive orientation (meaning that as you look down on the surface from any outward normal to the surface, the orientation of C is counterclockwise). Then

$$\int_C \mathbf{F} \cdot d\mathbf{r} = \iint_S (\text{curl } \mathbf{F}) \cdot d\mathbf{S}.$$

1) Use Stokes' Theorem to rewrite $\int_C \mathbf{F} \cdot d\mathbf{r}$ where $\mathbf{F}(x, y, z) = xy\mathbf{i} + yz\mathbf{j} + xz\mathbf{k}$ and C is the (counterclockwise) oriented triangular boundary of the intersection of the plane $x + 2y + 4z = 8$ with the first octant.

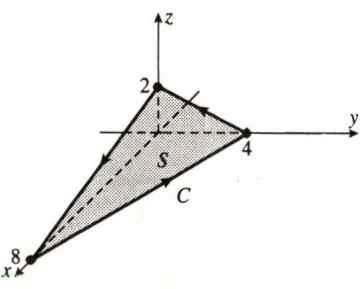

Let S be a plane surface bounded by C. By Stokes' Theorem:

$$\int_C \mathbf{F} \cdot d\mathbf{r} = \iint_S \text{curl } \mathbf{F} \cdot d\mathbf{S}.$$

$\text{curl } \mathbf{F} = (0 - y)\mathbf{i} + (0 - z)\mathbf{j} + (0 - x)\mathbf{k}$
$\qquad = -y\mathbf{i} - z\mathbf{j} - x\mathbf{k}.$

The surface S is

$$x = 8 - 2u - 4v, y = u, z = v$$

$$0 \le u \le 4, 0 \le v \le 2 - \frac{u}{2}.$$

$$\mathbf{r}_u \times \mathbf{r}_v = \begin{vmatrix} \mathbf{i} & \mathbf{j} & \mathbf{k} \\ -2 & 1 & 0 \\ -4 & 0 & 1 \end{vmatrix} = \mathbf{i} + 2\mathbf{j} + 4\mathbf{k}$$

is the normal vector to the plane.

$$\text{curl } \mathbf{F} \cdot (\mathbf{r}_u \times \mathbf{r}_v)$$
$$= -y(1) + (-z)2 + (-x)4$$
$$= -y - 2z - 4x$$
$$= -u - 2v - 4(8 - 2u - 4v)$$
$$= 7u + 14v - 32.$$

Thus $\displaystyle\iint_S \text{curl } \mathbf{F} \cdot d\mathbf{S}$

$$= \int_0^4 \int_0^{2-(u/2)} (7u + 14v - 32) dv \, du.$$

2) Use Stokes' Theorem to rewrite $\displaystyle\iint_S \text{curl } \mathbf{F} \cdot d\mathbf{S}$, where

$$\mathbf{F}(x, y, z) = (x - y)\mathbf{i} + (y - z)\mathbf{j} + (x - z)\mathbf{k}$$

and S is the paraboloid $z = 4 - x^2 - y^2$, $z \ge 0$, oriented upward.

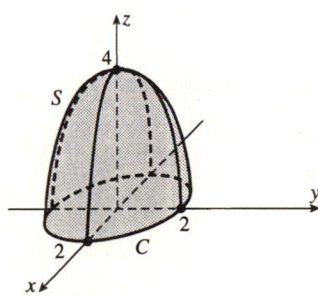

The boundary of S is the circle C in the xy plane $x^2 + y^2 = 4, z = 0$.

$$\mathbf{r}(t) = 2 \cos t\mathbf{i} + 2 \sin t\mathbf{j}, 0 \le t \le 2\pi.$$
$$\mathbf{r}'(t) = -2 \sin t\mathbf{i} + 2 \cos t\mathbf{j}.$$
$$\mathbf{F}(\mathbf{r}(t)) = 2((\cos t - \sin t)\mathbf{i} + \sin t\mathbf{j} + \cos t\mathbf{k})$$
$$\mathbf{F}(\mathbf{r}(t)) \cdot \mathbf{r}'(t) = 4(-\cos t \sin t + \sin^2 t$$
$$+ \cos t \sin t)$$
$$= 4 \sin^2 t.$$

Thus $\displaystyle\iint_S \text{curl } \mathbf{F} \cdot d\mathbf{S} = \iint_C \text{curl } \mathbf{F} \cdot d\mathbf{r}$

$$= \int_0^{2\pi} 4 \sin^2 t \, dt.$$

Section 17.9   The Divergence Theorem

Stokes' Theorem moves the double integrals and line integrals of Green's Theorem to three-dimensions. In this section, you will see that the Divergence Theorem may be thought of as boosting the integrals in Green's Theorem up by one dimension. Instead of Green's double integral relationship to a line integral, the Divergence Theorem relates a triple integral over a solid E in three-dimensions to a surface integral around the boundary of E.

Concepts to Master

Divergence Theorem

Summary and Focus Questions

Page 1163 (ET Page 1127)

Divergence Theorem: Let E be a simple solid whose boundary surface S has positive orientation. Let $\mathbf{F} = P\mathbf{i} + Q\mathbf{j} + R\mathbf{k}$, where P, Q, and R are component functions with continuous first partial derivatives on an open region containing E. Then

$$\iint_S \mathbf{F} \cdot d\mathbf{S} = \iiint_E \operatorname{div} \mathbf{F}\, dV.$$

1) Write a triple iterated integral for $\iint_S \mathbf{F} \cdot d\mathbf{S}$, where $\mathbf{F}(x, y, z) = xy\mathbf{i} + y^2\mathbf{k} + xz\mathbf{k}$ and S is the triangular surface bounded by $6x + 4y + 3z = 12$ in the first octant.

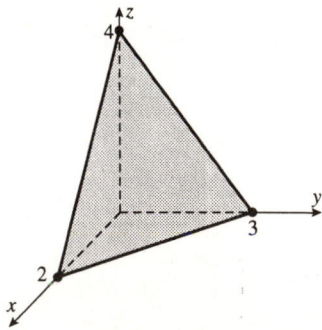

Let E be the solid region enclosed by S:
$$0 \le x \le 2$$
$$0 \le y \le 3 - \frac{3}{2}x$$
$$0 \le z \le 4 - 2x - \frac{4}{3}y.$$

$$\text{div } \mathbf{F} = \frac{\partial}{\partial x}(xy) + \frac{\partial}{\partial y}(y^2) + \frac{\partial}{\partial z}(xz)$$
$$= y + 2y + x = x + 3y.$$

Thus $\iint\limits_{S} \mathbf{F} \cdot d\mathbf{S} = \iiint\limits_{E}(x + 3y)dV$

$$= \int_0^2 \int_0^{3-(3/2)x} \int_0^{4 - 2x - (4/3)y}(x + 3y)dz\, dy\, dx.$$

2) Let $\mathbf{F} = (x, y, z) = \langle x, y^2, z^3 \rangle$.
   a) Find div $\mathbf{F}$.

$$\text{div } \mathbf{F} = 1 + 2y + 3z^2$$

   b) Write a surface integral for
   $$\iiint\limits_{E}(1 + 2y + 3z^2)dV,$$
   where E is the solid ball
   $x^2 + y^2 + z^2 \leq 1$ in the first octant.

In the first octant, the surface S of E may be parameterized by $\mathbf{r}(u, v)$:

$$x = \sin u \cos v$$
$$y = \sin u \sin v$$
$$z = \cos u,$$

$$0 \leq u \leq \frac{\pi}{2}, 0 \leq v \leq \frac{\pi}{2}.$$
$$\iiint\limits_{E}(1 + 2y + 3z^2)dV = \iint\limits_{S} \mathbf{F} \cdot d\mathbf{S}.$$

   c) Write an iterated integral for your answer to part b).

$$\mathbf{r}_u \times \mathbf{r}_v = \begin{vmatrix} \mathbf{i} & \mathbf{j} & \mathbf{k} \\ \cos u \cos v & \cos u \sin v & -\sin u \\ -\sin u \sin v & \sin u \cos v & 0 \end{vmatrix}$$

$$= (\sin^2 u \cos v)\mathbf{i} + (\sin^2 u \sin v)\mathbf{j}$$
$$+ (\sin u \cos u)\mathbf{k}$$

$|\mathbf{r}_u \times \mathbf{r}_v| =$
$$\sqrt{\sin^4 u \cos^2 v + \sin^4 u \sin^2 v + \sin^2 u \cos^2 u}$$

$$= \sin u.$$

Thus $\iint\limits_{S} \mathbf{F} \cdot d\mathbf{S}$

$$= \int_0^{\pi/2} \int_0^{\pi/2}(1 + 2\sin u \sin v$$
$$+ 3\cos^2 u)\sin u\, dv\, du.$$

Technology Plus for Chapter 17

1) Use a CAS to draw the vector field
$\mathbf{F}(x, y) = \langle 1 + x^2 + y^2, 2y(x + 1)\rangle$.

Does the field appear to be conservative?

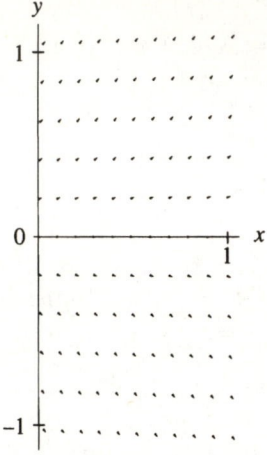

There is a chance that $\mathbf{F}$ is conservative because it appears that $\int_C \mathbf{F} \cdot d\mathbf{r}$ will be small (near zero) for a closed curve C. (The Riemann sum will contain positive and negative terms.)

For $P(x, y) = 1 + x^2 + y^2$ and
$P(x, y) = 2y(x + 1)$,

$\dfrac{\partial P}{\partial y} = 2y$ and $\dfrac{\partial Q}{\partial x} = 2y$.

Thus $\mathbf{F}$ is conservative.

2) a) Evaluate the line integral $\displaystyle\int_C \mathbf{F} \cdot d\mathbf{r}$,

where $\mathbf{F}(x, y) = xy\mathbf{i} + \dfrac{e^{x^2}}{4}\mathbf{j}$ and C is the

curve $\mathbf{r}(t) = t\mathbf{i} + t^2\mathbf{j},\ 0 \le t \le 2$

$\mathbf{F}(\mathbf{r}(t)) = t(t^2)\mathbf{i} + \dfrac{e^{t^2}}{4}\mathbf{j} = t^3\mathbf{i} + \dfrac{e^{t^2}}{4}\mathbf{j}.$

$\mathbf{r}'(t) = 1\mathbf{i} + 2t\mathbf{j}.$

$\mathbf{F}(\mathbf{r}(t)) \cdot \mathbf{r}'(t) = t^3 + \dfrac{te^{t^2}}{2}.$

$\displaystyle\int_C \mathbf{F} \cdot d\mathbf{r} = \int_0^2 \left(t^3 + \dfrac{te^{t^2}}{2}\right)dt$

$= \left(\dfrac{t^4}{4} + \dfrac{e^{t^2}}{4}\right)\Big]_0^2 = \dfrac{e^4 + 15}{4}.$

b) Illustrate part a) by graphing on the same screen C and $\mathbf{F}(\mathbf{r}(t))$ for $t = \frac{1}{2}, 1,$ and $\frac{3}{2}$.

$$\mathbf{F}\left(\mathbf{r}\left(\tfrac{1}{2}\right)\right) = \left\langle \tfrac{1}{8}, \tfrac{e^{1/4}}{4} \right\rangle \approx \langle 0.125, 0.321 \rangle.$$

$$\mathbf{F}(\mathbf{r}(1)) = \left\langle 1, \tfrac{e}{4} \right\rangle \approx \langle 1, 0.679 \rangle.$$

$$\mathbf{F}\left(\mathbf{r}\left(\tfrac{3}{2}\right)\right) = \left\langle \tfrac{27}{8}, \tfrac{e^{9/4}}{4} \right\rangle \approx \langle 3.375, 2.372 \rangle.$$

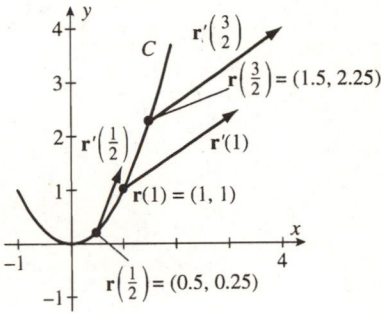

3) a) Use a CAS to sketch the graph of the surface

$$x = 4 + u \cos v + \frac{u^2}{4}$$

$$y = 4 + u \sin v + \frac{u^2}{4}$$

$$z = u$$

$$0 \le u \le 3, 0 \le v \le 2\pi.$$

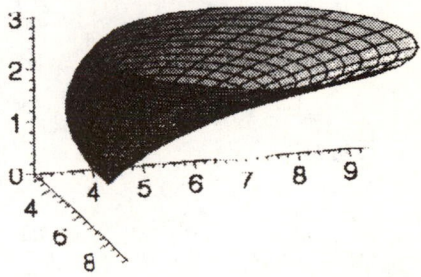

b) Set up an iterated integral for the surface area of the "tornado".

$$\mathbf{r}_u = \langle \cos v + \tfrac{u}{2}, \sin v + \tfrac{u}{2}, 1 \rangle$$

$$\mathbf{r}_v = \langle -u \sin v, u \cos v, 0 \rangle$$

$$\mathbf{r}_u \times \mathbf{r}_v = \langle -u \cos v, -u \sin v, u + \tfrac{u^2}{2}(\cos v + \sin v) \rangle.$$

$$|\mathbf{r}_u \times \mathbf{r}_v| = $$

$$\sqrt{2u^2 + u^3(\cos v + \sin v) + \tfrac{u^4}{4}(1 + 2\cos v \sin v)}.$$

The surface area is $\displaystyle \int_0^3 \int_0^{2\pi} |\mathbf{r}_u \times \mathbf{r}_v| \, du \, dv = $

$$\int_0^3 \int_0^{2\pi} \sqrt{2u^2 + u^3(\cos v + \sin v) + \tfrac{u^4}{4}(1 + 2\cos v \sin v)} \; dv \, du.$$

Chapter 18 — Second-Order Differential Equations

"THIS IS THE PART I ALWAYS HATE."

Cartoons courtesy of Sidney Harris. Used by permission.

Section 18.1   Second Order Linear Equations

This chapter is a continuation of chapter 10 on differential equations (chapter 9 in the ET edition) where you studied first-order linear differential equations. This section begins the extension of that study to second-order linear equations.

Concepts to Master

A. Second order linear equations; Homogeneous; Linear combination; Linearly independent; Auxiliary equation

B. Initial-values; Boundary-values

Summary and Focus Questions

Page 1177 (ET Page 1141)*

A. A *second order linear differential equation* has the form
$$P(x)y'' + Q(x)y' + R(x)y = G(x).$$
This section considers the *homogeneous* case, where $G(x) = 0$.

If y_1 and y_2 are two solutions to the homogeneous equation then all linear combinations $c_1y_1 + c_2y_2$ are also solutions (c_1, c_2 real numbers).

Solutions y_1 and y_2 are *linearly independent* means neither y_1 nor y_2 is a constant multiple of the other. If y_1 and y_2 are linearly independent solutions, then *all* solutions may be written as linear combinations of y_1 and y_2: $c_1y_1 + c_2y_2$.

In the case where P, Q, and R are constants, the equation has the form $ay'' + by' + cy = 0$. The corresponding *auxiliary equation* (with variable r) is $ar^2 + br + c = 0$. The type of roots (real or imaginary) of the auxiliary equation determine the type of roots to $ay'' + by' + cy = 0$:

Discriminant	Roots	Solution
$b^2 - 4ac > 0$	r_1, r_2 real, $r_1 \neq r_2$	$y = c_1 e^{r_1 x} + c_2 e^{r_2 x}$
$b^2 - 4ac = 0$	r real	$y = c_1 e^{rx} + c_2 x e^{rx}$
$b^2 - 4ac < 0$	r_1, r_2 complex, $r_1 \neq r_2$ $r_1 = \alpha + \beta i$ $r_2 = \alpha - \beta i$	$y = e^{\alpha x}(c_1 \cos \beta x + c_2 \sin \beta x)$

1) True or False:
$xy' + y'' + 5x + 10y = 0$ is homogeneous.

False (because of the term $5x$).

2) True or False:
$y_1 = 2xy - 3y^2$ and $y_2 = 12y^2 - 8xy$ are linearly independent.

False, $-4y_1 = y_2$, so they are dependent.

3) Sometimes, Always, or Never:
If y_1 and y_2 are solutions to
$ay'' + by' + cy = 0$ then all solutions are of the form $c_1y_1 + c_2y_2$.

Sometimes; y_2 and y_1 must be linearly independent for this to be true.

4) Find the general solution to each:

a) $y'' + 6y' + 8y = 0$

The auxiliary equation is $r^2 + 6r + 8 = 0$.
$(r + 2)(r + 4) = 0$.
Thus $r = -2, -4$.
The general solution is $y = c_1e^{-2x} + c_2e^{-4x}$.

b) $y'' - 6y' + 13y = 0$

The auxiliary equation is $r^2 - 6r + 13 = 0$.
$$r = \frac{-(-6) \pm \sqrt{36 - 4(13)}}{2} = \frac{6 \pm \sqrt{-16}}{2}$$
$$= 3 \pm 2i.$$
The general solution is
$y = e^{3x}(c_1 \cos 2x + c_2 \sin 2x)$.

c) $4y'' - 12y' + 9y = 0$

The auxiliary equation is
$4r^2 - 12r + 9 = 0$.
$(2r + 3)^2 = 0$.
Thus $r = \frac{3}{2}$.
The general solution is
$y = c_1e^{(3/2)x} + c_2xe^{(3/2)x}$.

Page
1181
(ET Page
1145)

B. To specify a particular solution of a second order equation two conditions must be given. Depending on what is specified we have two classes of problems.

1) If $y(x_0) = y_0$ and $y'(x_0) = y_1$ are specified, we have an *initial-value problem*.

2) If $y(x_0) = y_0$ and $y(x_1) = y_1$ are specified, we have an *boundary-value problem*.

For continuous P, Q, R, and G with $P \neq 0$, initial-value problems will have a solution but boundary-value problems need not necessarily have a solution.. The method for determining the particular solution involves substituting the conditions in y and y' and solving two equations in the two unknowns c_1 and c_2.

5) Solve $y'' + 2y' - 8y = 0$ with initial conditions $y(0) = 28$ and $y'(0) = 2$.

$r^2 + 2r - 8 = 0$ has solutions $r = -4, 2$.
The general solution is $y = c_1 e^{2x} + c_2 e^{-4x}$.
Since $y(0) = 28$, $c_1 + c_2 = 28$.
$y' = 2c_1 e^{2x} - 4c_2 e^{-4x}$.
Since $y'(0) = 2$, $2c_1 - 4c_2 = 2$
or $c_1 - 2c_2 = 1$.
The system $c_1 + c_2 = 28$
$\qquad\qquad c_1 - 2c_2 = 1$
has solution $3c_2 = 27$
$\qquad\qquad c_2 = 9$.
Thus $c_1 - 18 = 1$, $c_1 = 19$.
The particular solution is
$y = 19 e^{2x} + 9 e^{-4x}$.

6) Solve $y'' - 4y' + 3y = 0$ with boundary conditions $y(0) = e^2$, $y(1) = e$.

$r^2 - 4r + 3 = 0$
$(r - 3)(r - 1) = 0$
$r = 1, 3$.

The general solution is
$y = c_1 e^x + c_2 e^{3x}$.
$y(0) = e^2$ implies $c_1 + c_2 = e^2$.
$y(1) = e$ implies $c_1 e + c_2 e^3 = e$,
or $c_1 + c_2 e^2 = 1$.

Subtract the second equation from the first:
$c_2 - c_2 e^2 = e^2 - 1$
$c_2(1 - e^2) = e^2 - 1$
$c_2(1 - e^2) = -(1 - e^2)$
$c_2 = -1$.

Then $c_1 - 1 = e^2$ and so $c_1 = e^2 + 1$.
The solution is $y = (e^2 + 1)e^x - e^{3x}$.

7) Solve $y'' - 6y' + 9y = 0$ with boundary conditions $y(0) = 2, y(1) = e^3$.

$r^2 - 6r + 9 = 0$ has solution $r = 3$.
The general solution is $y = c_1 e^{3x} + c_2 x e^{3x}$.

$y(0) = 2$ implies $c_1 = 2$.
$y(1) = 1$ implies $c_1 e^3 + c_2 e^3 = e^3$.

$c_1 + c_2 = 1$. Since $c_1 = 2, c_2 = -1$.
Thus $y = 2e^{3x} - xe^{3x}$.

Section 18.2 Nonhomogeneous Linear Equations

The previous section discussed methods for solving second-order homogeneous equations of the form $ay'' + by' + cy = 0$. This section considers nonhomogeneous equations of the form $ay'' + by' + cy = G(x)$ and will relate their solutions back to the homogeneous case.

Concepts to Master

A. Nonhomogeneous linear differential equation; Method of Undetermined Coefficients

B. Method of Variation of Parameters

Summary and Focus Questions

Page 1184 (ET Page 1148)

A. A second order *linear nonhomogeneous differential equation* with constant differential coefficients has the form $ay'' + by' + cy = G(x)$, where $G(x)$ is continuous. The corresponding *complementary equation* is
$$ay'' + by' + cy = 0.$$

The general solution is $y(x) = y_p(x) + y_c(x)$ where y_c is the general solution of the complementary equation and y_p is a particular solution of the nonhomogeneous equation.

The particular solution y_p can sometimes be found using the *method of undetermined coefficients*. The solution $y_p(x)$ is given in the following table and depends on the form of $G(x)$.

$G(x)$ form	Possible $y_p(x)$ form
polynomial: $C_n x^n + \cdots + C_1 x + C_0$	$A_x x^n + \cdots + A_1 x + A_0$
exponential: Ce^{kx}	$A_1 x^r e^{kx}$, where $r = 0$, 1, or 2
trigonometric: $C \cos kx + D \sin kx$	$A_1 x^r \cos kx + A_2 x^r \sin kx$, where $r = 0$ or 1.

Note: In the last two cases r is the smallest nonnegative integer such that $y_p(x)$ is *not* a solution to the complementary equation. First try $r = 0$. If any term of $y_p(x)$ is a solution to the complementary equation, then try $r = 1$ or if necessary, $r = 2$.

We then substitute $y_p(x)$ and its derivatives into the nonhomogeneous equation to obtain a system of equations to determine $A_1, A_2, \ldots$.

1) Solve the equation $y'' + 3y' - 10y = 16e^{3x}$.

The complementary equation
$y'' + 3y' - 10y = 0$ has auxiliary equation
$r^2 + 3r - 10 = 0$. $(r - 2)(r + 5) = 0$, so
$r = 2, -5$. The general solution to the
nonhomogeneous equation is
$y(x) = c_1 e^{2x} + c_2 e^{-5x} + y_p(x)$.

To find $y_p(x)$ we note the form of $G(x)$ is
$16e^{3x}$ so $y_p(x) = Ax^r e^{3x}$ is a good trial form.
First consider $r = 0$, $y_p(x) = Ae^{3x}$.
$y_p'(x) = 3Ae^{3x}$, $y_p''(x) = 9Ae^{3x}$.

Substituting these into the original equation
yields $9Ae^{3x} + 3(3Ae^{3x}) - 10(Ae^{3x}) = 16e^{3x}$
$8e^{3x} = 16e^{3x}$, $A = 2$.

Hence $y_p = 2e^{3x}$. (We note $2e^{3x}$ does not
satisfy the complementary equation so
$r = 0$ is valid.) The general solution is
$y(x) = c_1 e^{2x} + c_2 e^{-5x} + 2e^{3x}$.

2) Find a particular solution to
$y'' - 7y' + 3y = 6x - 2$.

The form $6x - 2$ suggests
$y_p(x) = Ax + B$. $y_p'(x) = A$, $y_p''(x) = 0$.

Substituting into the original equation:
$0 - 7(A) + 3(Ax + B) = 6x - 2$
$3Ax + (-7A + 3B) = 6x - 2$

Comparing coefficients
$3A = 6$ and $-7A + 3B = -2$.

Thus $A = 2$ and $-7(2) + 3B = -2$,
$B = 4$. Therefore $y_p(x) = 2x + 4$.

Page 1188
(ET Page 1152)

B. The method of *variation of parameters* given below, unlike the undetermined
coefficients method, does not rely on the form of $G(x)$ to find a particular
solution $y_p(x)$ to $ay'' + by' + cy = G(x)$:

1) Find linearly independent solutions y_1 and y_2 to $ay'' + by' + cy = 0$.
2) Solve the following system for u_1' and u_2':
$$u_1'y_1 + u_2'y_2 = 0$$
$$a(u_1'y_1 + u_2'y_2') = G(x).$$
3) Integrate u_1' and u_2' to get $u_1(x)$ and $u_2(x)$.
4) A particular solution is $y_p(x) = u_1(x)y_1(x) + u_2(x)y_2(x)$.

3) Solve $y'' - 5y' + 4y = 30e^{6x}$.

1) The auxiliary equation is
   $y'' - 5y' + 4y = 0$.
   From $r^2 - 5r + 4 = 0$,
   $(r - 1)(r - 4) = 0, r = 1, r = 4$.
   Thus the general solution to
   $y'' - 5y' + 4y = 0$ is $y_c(x) = c_1e^x + c_2e^{4x}$.
   Let $c_1 = 1, c_2 = 0$, then $y_1 = e^x$.
   Let $c_1 = 0, c_2 = 1$, then $y_2 = e^{4x}$.

   y_1 and y_2 are linearly independent and
   $y_1' = e^x, y_2' = 4e^{4x}$.

2) We solve the system
   $u_1'e^x + u_2'e^{4x} = 0$
   $u_1'e^x + u_2'(4e^{4x}) = 30e^{6x}$.

   From the first equation, $u_1' = -u_2'e^{3x}$.
   Substituting into the second,
   $(-u_2'e^{3x})e^x + u_2'4e^{4x} = 30e^{6x}$
   $3u_2'e^{4x} = 30e^{6x}$
   $u_2' = 10e^{2x}$.
   Thus $u_1' = -(10e^{2x})e^{3x} = -10e^{5x}$.

3) From $u_1' = -10e^{5x}, u_1 = -2e^{5x}$ and from
   $u_2' = 10e^{2x}, u_2 = 5e^{2x}$.

4) The particular solution is
   $y_p(x) = u_1y_1 + u_2y_2$
   $= (-2e^{5x})e^x + (5e^{2x})e^{4x} = 3e^{6x}$.

   Therefore the general solution to the
   nonhomogeneous equation is
   $y(x) = c_1e^x + c_2e^{4x} + 3e^{6x}$.

Section 18.3   Applications of Second-Order Differential Equations

Second-order differential equations have many applications in science and engineering. This section presents two of them – current within circuits and damped oscillating motion.

Concepts to Master

Models of current in circuits and damped motion

Summary and Focus Questions

Page 1194 (ET Page 1158)

The equation $ay'' + by' + cy = G(x)$ has applications in several areas including

1) Damped motion: $mx'' + cx' + kx = F(t)$
   The interpretation is

   m = mass of an object at the end of a spring
   x = displacement from equilibrium at time t
   x', x'' = velocity and acceleration
   c = damping constant (such as air resistance)
   k = spring constant
   $F(t)$ = external force (such as gravity)

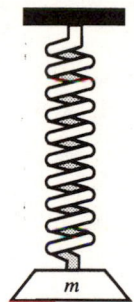

2) Electric circuits: $LQ'' + RQ' + \dfrac{1}{C}Q = E(t)$
   The interpretation is

   Q = charge on the capacitor at time t
   Q' = current
   L = inductance constant for an inductor
   R = resistance constant for a resistor
   $\dfrac{1}{C}$ = elastance constant for a capacitor
   $E(t)$ = electromotive force applied to a circuit

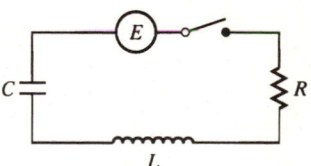

1) A spring with a 10-kg mass can be stretched 0.5 m beyond its equilibrium by a force of 85 N (newtons). Suppose the spring is held in a fluid with damping constant 20. If the mass starts at equilibrium with an initial velocity of 1 m/sec, find the position of the mass after t seconds.

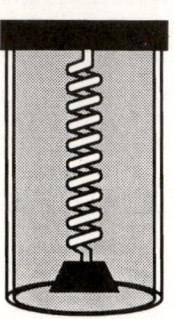

To find the spring constant, k, use Hooke's law: $k(0.5) = 85$, $k = 170$.
The differential equation model is
$10x'' + 20x' + 170x = 0$, or
$x'' + 2x' + 17x = 0$
with initial values $x(0) = 0$, $x'(0) = 1$.
From $r^2 + 2r + 17 = 0$, $r = -1 \pm 4i$.

Thus the general solution is
$x = e^{-t}(c_1 \cos 4t + c_2 \sin 4t)$.
At $t = 0$, $x = 0$, so
$0 = 1(c_1 \cos 0 + c_2 \sin 0)$.
Therefore $c_1 = 0$.
Thus $x = e^{-t}(c_2 \sin 4t)$.
$x' = e^{-t}(4c_2 \cos 4t) - e^{-t}(c_2 \sin 4t)$.

At $t = 0$, $x' = 1$, so
$1 = 1(4c_2) - 1(c_2(0))$, $c_2 = 0.25$.
Thus $x = 0.25e^{-t} \sin 4t$ is the position after t seconds.

2) A series circuit consists of a resistor with $R = 40\Omega$ (ohms), an inductor with $L = 1$ H (henries), a capacitor with $C = 0.01$ F (farads), and a generator producing a voltage of $3 + 2 \sin t$. Find the differential equation for determining the charge at time t.

$L = 1$, $R = 40$, $C = 0.01$, $E(t) = 3 + 2 \sin t$.
The equation is
$Q'' + 40Q' + 100Q = 3 + 2 \sin t$.

Section 18.4 Series Solutions

Some second-order equations, even some simple looking ones, cannot be solved explicitly using combinations of our familiar functions. Thus this section shows you how to solve some equations by representing the solution as a power series.

Concepts of Master

Power series solution to a differentiation equation

Summary and Focus Questions

The method of solving a differential equation using a power series has these steps:

Page 1199 (ET Page 1163)

1) Assume the equation has a solution of the form

$$y = \sum_{n=0}^{\infty} c_n x^n = c_0 + c_1 x + c_2 x^2 + c_3 x^3 + \cdots$$

2) Obtain expressions for y', y'', etc.

$$y' = \sum_{n=1}^{\infty} n c_n x^{n-1} = c_1 + 2c_2 x + 3c_3 x^2 + \cdots$$

$$y'' = \sum_{n=2}^{\infty} n(n-1) c_n x^{n-2} = 2c_2 + 3 \cdot 2c_3 x + 4 \cdot 3c_4 x^2 + \cdots$$

3) Substitute these expressions into the differential equation.

4) Equate coefficients of corresponding powers of x to obtain equations to determine $c_0, c_1, c_2, \cdots$. (The general expression for c_i will often be a recursive expression in terms of previous c_j, where $j < i$.)

1) Find a series solution to $y'' = xy'$.

1) $y = c_0 + c_1 x + c_2 x^2 + c_3 x^3 + \cdots$,

2) $y' = c_1 + 2c_2 x + 3c_3 x^2 + 4c_4 x^3 + \cdots$, and
$y'' = 2c_2 + 3 \cdot 2c_3 x + 4 \cdot 3c_4 x^2 + 5 \cdot 4c_5 x^3 + \cdots$.

3) Thus $xy' = c_1 x + 2c_2 x^2 + 3c_2 x^3 + \cdots$.

4) From the equation $y'' = xy'$ we equate coefficients.

Power of x	Relation
x^0	$2c_2 = 0$
x^1	$3 \cdot 2c_3 = c_1$
x^2	$4 \cdot 3c_4 = 2c_2$
x^3	$5 \cdot 4c_5 = 3c_3$

In general, $n(n - 1)c_n = (n - 2)c_{n-2}$.

$$c_n = \frac{(n - 2)c_{n-2}}{n(n - 1)}$$

Thus c_0 is arbitrary, c_1 is arbitrary, $c_2 = 0$,

$$c_3 = \frac{c_1}{3 \cdot 2}, \quad c_4 = 0,$$

$$c_5 = \frac{3c_3}{5 \cdot 4} = \frac{3c_1}{5 \cdot 4 \cdot 3 \cdot 2}, \quad c_6 = 0,$$

$$c_7 = \frac{5c_5}{7 \cdot 6} = \frac{5 \cdot 3c_1}{7 \cdot 6 \cdot 5 \cdot 4 \cdot 3 \cdot 2}, \text{ etc.}$$

In general

$$c_n = \begin{cases} 0 & n \text{ even} \\ \frac{(n - 2)(n - 4) \cdots 1}{n!}c_1 & n \text{ odd} \end{cases}$$

or equivalently, $c_{2n} = 0$, and

$$c_{2n + 1} = \frac{(2n - 1) \cdots 3 \cdot 1}{n!}c_1 \text{ for all } n.$$

Thus $y = c_0 + c_1 \sum_{n = 1}^{\infty} \frac{(2n - 1) \cdots 3 \cdot 1}{(2n + 1)!} x^{2n+1}$

is the solution to $y'' = xy'$, where c_0 and c_1 are arbitrary constants.

2) Find a series solution to $y'' + xy' - y = 0$, $y(0) = 1, y'(0) = 0$.

1) $y = \sum_{n = 0}^{\infty} c_n x^n$

2) $y' = \sum_{n = 1}^{\infty} nc_n x^{n-1}$

$$xy' = \sum_{n = 1}^{\infty} nc_n x^n = \sum_{n = 0}^{\infty} nc_n x^n.$$

$$y'' = \sum_{n = 2}^{\infty} n(n - 1)c_n x^{n-2}$$

$$= \sum_{n = 0}^{\infty} (n + 2)(n + 1)c_{n+2} x^n.$$

3) $\sum_{n = 0}^{\infty} (n + 2)(n + 1)c_{n+2} x^n + \sum_{n = 0}^{\infty} nc_n x^n$

$$- \sum_{n = 0}^{\infty} c_n x^n = 0.$$

$$\sum_{n = 0}^{\infty} [(n + 2)(n + 1)c_{n+2} + (n - 1)c_n]x^n = 0.$$

4) $(n + 2)(n + 1)c_{n+2} + (n - 1)c_n = 0$ for all n.

Thus $c_{n+2} = \frac{-(n - 1)}{(n + 2)(n + 1)}c_n$.

$n = 0$: $c_2 = \frac{1}{2}c_0$

$n = 1$: $c_3 = 0$

$n = 2$: $c_4 = \frac{-1}{4 \cdot 3} c_2 = \frac{-1}{4 \cdot 3 \cdot 2} c_0$

$n = 3$: $c_5 = 0$

$n = 4$: $c_6 = \dfrac{-3}{6 \cdot 5} c_4 = \dfrac{-3}{6!} c_0$

$n = 5$: $c_7 = 0$

$n = 6$: $c_8 = \dfrac{(-5)(-3)}{8!} c_0$

and so on.

Since $y(0) = 1$, $c_0 = 1$.

Since $y'(0) = 0$, $c_1 = 0$.

For $n \geq 2$, $c_n = \dfrac{(1 - n)(3 - n) \ldots (-1)(1)}{n!}$ if

n is even,

and $c_n = 0$ if n is odd.

Thus

$$y = 1 + \sum_{n=1}^{\infty} \frac{(3 - 2n) \ldots (-3)(-1)(1)}{2n!} x^{2n}.$$

Technology Plus for Chapter 18

1) **a)** Graph the two basic solutions $y = e^{r_1 x}$
and $y = e^{r_2 x}$ for $12y'' + y' - y = 0$.

$$12x^2 + x - 1 = 0$$
$$(3x + 1)(4x - 1) = 0$$
$$x = -\frac{1}{3} \text{ and } x = \frac{1}{4}.$$

The basic solutions are

$$y = e^{-x/3} \text{ and } y = e^{x/4}.$$

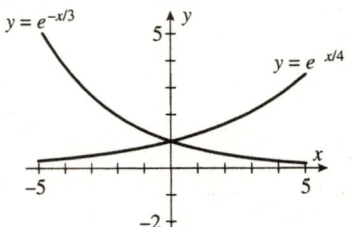

b) Now graph on the same screen
$y = c_1 e^{r_1 x} + c_2 e^{r_2 x}$ for several values of
c_1 and c_2, such as $c_1 = \pm 1, \pm 2$ and
$c_2 = \pm 1, \pm 2$.

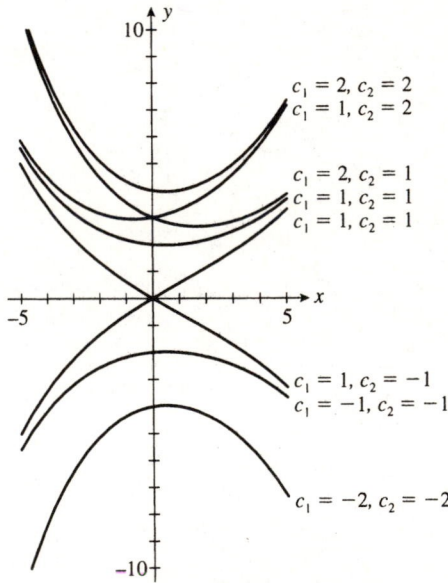

c) What features do all the solutions have
in common?

For any c_1 and c_2, not both zero,

$$\lim_{x \to \infty} y = \pm\infty \text{ and } \lim_{x \to -\infty} y = \pm\infty. \text{ All}$$

solutions cross the y-axis at $c_1 + c_2$.

Section 11.1

_____ **1.** The graph of $x = 2 + 3t,\ y = 4 - t$ is a:

    a) circle   b) ellipse

    c) line   d) parabola

_____ **2.** The best graph of $x = \cos t,\ y = \sin^2 t$ is:

a)

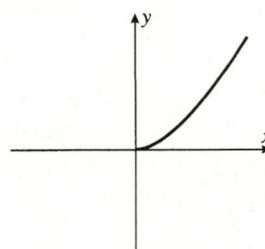

b)

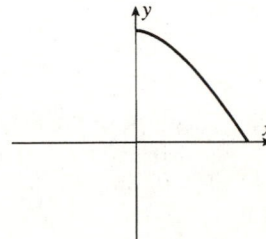

c)

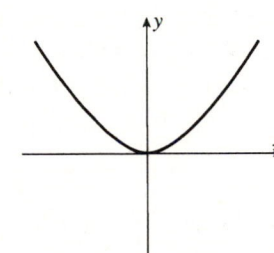

d)

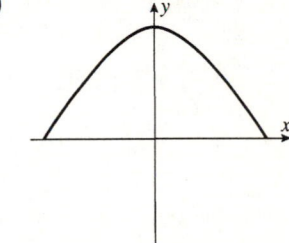

_____ **3.** Elimination of the parameter in $x = 2t^{3/2},\ y = t^{2/3}$ gives:

    a) $x^4 = 16y^9$ b) $16x^4 = y^4$

    c) $x^3 = 8y^4$ d) $8x^3 = y^4$

Section 11.2

_____ 1. Find $\frac{dy}{dx}$ if $x = \sqrt{t}$, $y = \sin 2t$.

   a) $\frac{4\cos t}{\sqrt{t}}$ b) $\frac{\cos 2t}{\sqrt{t}}$ c) $\frac{\cos 2t}{2\sqrt{t}}$ d) $4\sqrt{t}\cos 2t$

_____ 2. Find $\frac{d^2y}{dx^2}$ for $x = 3t^2 + 1$, $y = t^6 + 6t^5$.

   a) $t^4 + 5t^3$ b) $4t^3 + 15t^2$

   c) $\frac{2}{3}t^2 + \frac{5}{2}t$ d) $t^3 + \frac{1}{2}t^2$

_____ 3. The slope of the tangent line at the point where $t = \frac{\pi}{6}$ to the curve $y = \sin 2t$, $x = \cos 3t$ is:

   a) $\frac{1}{3}$ b) $-\frac{1}{3}$ c) 3 d) -3

_____ 4. A definite integral for the area under the curve described by $x = t^2 + 1$, $y = 2t$, $0 \le t \le 1$ is:

   a) $\int_0^1 (2t^3 + 2t)\,dt$ b) $\int_0^1 4t^2\,dt$

   c) $\int_0^1 (2t^2 + 2)\,dt$ d) $\int_0^1 4t\,dt$

_____ 5. The length of the curve given by $x = 3t^2 + 2$, $y = 2t^3$, $t \in [0, 1]$ is:

   a) $4\sqrt{2} - 2$ b) $8\sqrt{2} - 1$

   c) $\frac{2}{3}\left(2\sqrt{2} - 1\right)$ d) $\sqrt{2} - 1$

_____ 6. Find a definite integral for the area of the surface of revolution about the x-axis obtained by rotating the curve $y = t^2$, $x = 1 + 3t$, $0 \le t \le 2$.

   a) $\int_0^2 2\pi t^2 \sqrt{t^4 + 9t^2 + 6t + 1}\,dt$

   b) $\int_0^2 2\pi t^2 \sqrt{4t^2 + 9}\,dt$

   c) $\int_0^2 2\pi(2t)\sqrt{t^4 + 9t^2 + 6t + 1}\,dt$

   d) $\int_0^2 2\pi(2t)\sqrt{4t^4 + 9}\,dt$

Section 11.3

1. A set of polar coordinates of the point P plotted at the right is:

   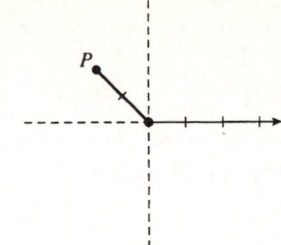

   a) $\left(-2, \frac{\pi}{4}\right)$

   b) $\left(-2, \frac{3\pi}{4}\right)$

   c) $\left(2, -\frac{\pi}{4}\right)$

   d) $\left(2, \frac{3\pi}{4}\right)$

2. A set of polar coordinates of the point with rectangular coordinates $(5, 5)$ is:

   a) $(25, 0)$ b) $\left(5, \frac{\pi}{4}\right)$

   c) $\left(5\sqrt{2}, \frac{\pi}{4}\right)$ d) $\left(50, -\frac{\pi}{4}\right)$

3. Rectangular coordinates of the point with polar coordinates $\left(-1, \frac{3\pi}{2}\right)$ are:

   a) $(-1, 0)$ b) $(0, 1)$

   c) $(0, -1)$ d) $(1, 0)$

4. The graph of $\theta = 2$ in polar coordinates is a:

   a) circle   b) line   c) spiral   d) 3-leaved rose

5. The slope of the tangent line to $r = \cos \theta$ at $\theta = \frac{\pi}{3}$ is:

   a) $\sqrt{3}$ b) $\frac{1}{\sqrt{3}}$ c) $-\sqrt{3}$ d) $-\frac{1}{\sqrt{3}}$

Section 11.4

_____ **1.** The area of the region bounded by $\theta = \frac{\pi}{3}$, $\theta = \frac{\pi}{4}$, and $r = \sec\theta$ is:

a) $\frac{1}{2}\left(\sqrt{3} - 1\right)$ b) $\sqrt{3}$

c) $2(\sqrt{3} - 1)$ d) $2\sqrt{3}$

_____ **2.** The area of the shaded region is given by:

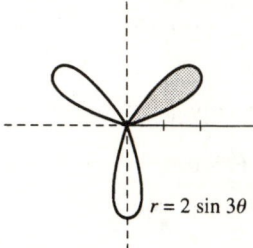

$r = 2\sin 3\theta$

a) $\int_0^{\pi/2} \sin 3\theta \, d\theta$

b) $\int_0^{\pi/2} 2\sin^2 3\theta \, d\theta$

c) $\int_0^{\pi/3} \sin 3\theta \, d\theta$

d) $\int_0^{\pi/3} 2\sin^2 3\theta \, d\theta$

_____ **3.** The length of the arc $r = e^\theta$ for $0 \le \theta \le \pi$ is:

a) $e^\pi - 1$ b) $2(e^\pi - 1)$

c) $\sqrt{2}(e^\pi - 1)$ d) $2\sqrt{2}(e^\pi - 1)$

Section 11.5

_____ **1.** The best graph of $\frac{y^2}{16} = 1 + \frac{x^2}{25}$ is:

a)

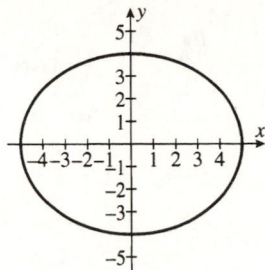

b)

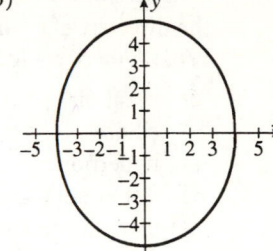

c)

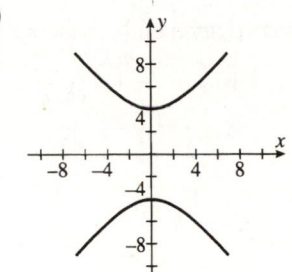

d)

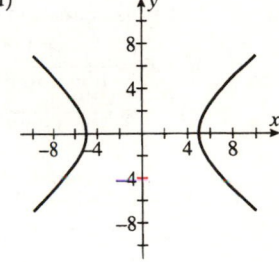

_____ **2.** The conic section whose equation is
$y(3 - y) + 4x^2 = 2x(1 + 2x) - y$ is a

a) parabola   b) ellipse   c) hyperbola   d) None of these

Section 11.6

_____ **1.** The figure at the right shows one point P on a conic and the distances of P to the focus and the directrix of a conic. What type of conic is it?

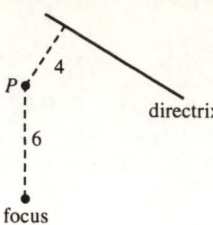

  a) parabola
  b) ellipse
  c) hyperbola
  d) not enough information is provided to answer the question

_____ **2.** The polar equation of the conic with eccentricity 3 and directrix $x = -7$ is:

  a) $r = \dfrac{21}{1 + 3\cos\theta}$ b) $r = \dfrac{21}{1 - 3\cos\theta}$

  c) $r = \dfrac{21}{1 + 3\sin\theta}$ d) $r = \dfrac{21}{1 - 3\sin\theta}$

_____ **3.** The directrix of the conic given by $r = \dfrac{6}{2 + 10\sin\theta}$ is:

  a) $x = \dfrac{5}{3}$ b) $x = \dfrac{3}{5}$ c) $y = \dfrac{5}{3}$ d) $y = \dfrac{3}{5}$

_____ **4.** The polar form for the graph at the right is:

  a) $r = \dfrac{ed}{1 + e\cos\theta}$

  b) $r = \dfrac{ed}{1 - 3\cos\theta}$

  c) $r = \dfrac{ed}{1 + e\sin\theta}$

  d) $r = \dfrac{ed}{1 - e\sin\theta}$

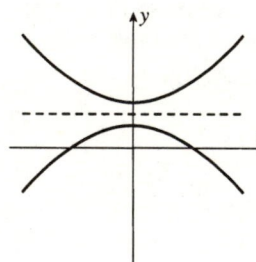

Section 12.1

_____ **1.** $\lim\limits_{n\to\infty} \dfrac{n^2 + 3n}{2n^2 + n + 1} =$

    a) 0   b) $\dfrac{1}{2}$ c) 1   d) ∞

_____ **2.** Sometimes, Always, or Never:
If $a_n \geq b_n \geq 0$ and $\{b_n\}$ diverges, then $\{a_n\}$ diverges.

_____ **3.** Sometimes, Always, or Never:
If $a_n \geq b_n \geq c_n$ and both $\{a_n\}$ and $\{c_n\}$ converge, then $\{b_n\}$ converges.

_____ **4.** Sometimes, Always, or Never:
If $\{a_n\}$ is increasing and bounded above, then $\{a_n\}$ converges.

_____ **5.** True, False:
$a_n = \dfrac{(-1)^n}{n^2}$ is monotonic.

_____ **6.** $\lim\limits_{n\to\infty} \dfrac{\arctan n}{2} =$

    a) $\dfrac{\pi}{4}$ b) $\dfrac{\pi}{2}$ c) π d) does not exist

_____ **7.** The fourth term of $\{a_n\}$ defined by $a_1 = 3$, and $a_{n+1} = \dfrac{2}{3}a_n$, $n = 1, 2, 3, \dots$, is:

    a) $\dfrac{16}{81}$ b $\dfrac{16}{27}$ c) $\dfrac{8}{27}$ d) $\dfrac{8}{9}$

Section 12.2

_____ **1.** True, False:

$$\sum_{n=1}^{\infty} a_n \text{ converges if } \lim_{n \to \infty} a_n = 0.$$

_____ **2.** True, False:

If $\sum_{n=1}^{\infty} a_n$ converges and $\sum_{n=1}^{\infty} b_n$ converges, then $\sum_{n=1}^{\infty} (a_n - b_n)$ converges.

_____ **3.** The harmonic series is:

a) $1 + 2 + 3 + 4 + \ldots$

b) $1 + \frac{1}{2} + \frac{1}{3} + \frac{1}{4} + \ldots$

c) $1 + \frac{1}{2} + \frac{1}{4} + \frac{1}{8} + \ldots$

d) $1 - \frac{1}{2} + \frac{1}{4} - \frac{1}{8} + \ldots$

_____ **4.** True, False:

$$\sum_{n=1}^{\infty} \frac{n}{n+1} \text{ converges.}$$

_____ **5.** $\sum_{n=1}^{\infty} 2(\frac{1}{4})^n$ converges to:

a) $\frac{9}{4}$ b) 2 c) $\frac{2}{3}$ d) the series diverges

_____ **6.** True, False:

$-3 + 1 - \frac{1}{3} + \frac{1}{9} - \frac{1}{27} + \ldots$ is a geometric series.

Section 12.3

_____ 1. For what values of p does the series $\sum\limits_{n=1}^{\infty} \dfrac{1}{(n^2)^p}$ converge?

    a) $p > -\dfrac{1}{2}$ b) $p < -\dfrac{1}{2}$ c) $p > \dfrac{1}{2}$ d) $p < \dfrac{1}{2}$

_____ 2. True, False:

    If $f(x)$ is continuous and decreasing, $f(n) = a_n$ for all $n = 1, 2, 3, \ldots$,

    and $\int_{1}^{\infty} f(x)\,dx = M$, then $\sum\limits_{n=1}^{\infty} a_n = M$.

_____ 3. Does $\sum\limits_{n=1}^{\infty} \dfrac{1}{n^{2/3}}$ converge?

_____ 4. Does $\sum\limits_{n=1}^{\infty} \dfrac{n + 2}{(n^2 + 4n + 1)^2}$ converge?

_____ 5. For $s = \sum\limits_{n=1}^{\infty} \dfrac{1}{n^3}$, an upper bound estimate for $s - s_6$ (where s_6 is the sixth partial sum) is:

    a) $\int_{1}^{\infty} x^{-3}\,dx$ b) $\int_{6}^{\infty} x^{-3}\,dx$

    c) $\int_{7}^{\infty} x^{-3}\,dx$ d)   the series does not converge

_____ **1.** Sometimes, Always, or Never:

If $0 \leq a_n \leq b_n$ for all n and $\displaystyle\sum_{n=1}^{\infty} a_n$ diverges, then $\displaystyle\sum_{n=1}^{\infty} b_n$ converges.

_____ **2.** Does $\displaystyle\sum_{n=1}^{\infty} \frac{n+1}{n^3}$ converge?

_____ **3.** Does $\displaystyle\sum_{n=1}^{\infty} \frac{\sqrt{n} + \sqrt[3]{n}}{n^{2/3} + n^{3/2} + 1}$ converge?

_____ **4.** Does $\displaystyle\sum_{n=1}^{\infty} \frac{\cos^2(2^n)}{2^n}$ converge?

_____ **5.** $s = \displaystyle\sum_{n=1}^{\infty} \frac{1}{n\,2^n}$ converges by the Comparison Test, comparing it to $\displaystyle\sum_{n=1}^{\infty} \frac{1}{2^n}$.

Using this information, make an estimate of the difference between s and its third partial sum.

a) $\dfrac{1}{16}$ b) $\dfrac{1}{8}$ c) $\dfrac{1}{32}$ d) 1

Section 12.5

_____ **1.** Does $\displaystyle\sum_{n=1}^{\infty} \frac{(-1)^{n+1} \ln n}{n^2}$ converge?

_____ **2.** Does $\displaystyle\sum_{n=1}^{\infty} \frac{(-1)^n}{\sqrt[4]{n+1}}$ converge?

_____ **3.** For what value of n is the nth partial sum within 0.01 of the value of $\displaystyle\sum_{n=1}^{\infty} \frac{(-1)^n}{2^n}$? (Choose the smallest such n.)

    a) $n = 4$ b) $n = 6$ c) $n = 8$ d) $n = 10$

_____ **4.** True, False:

The Alternating Series Test may be applied to determine the convergence of $\displaystyle\sum_{n=1}^{\infty} \frac{2 + (-1)^n}{2n^2}$.

Section 12.6

_____ **1.** True, False:

If $\displaystyle\sum_{n=1}^{\infty} a_n$ converges absolutely, then it converges conditionally.

_____ **2.** True, False:

If $\displaystyle\lim_{n\to\infty}\left|\frac{a_n}{a_{n+1}}\right| = 3$, then $\displaystyle\sum_{n=1}^{\infty} a_n$ converge absolutely.

_____ **3.** True, False:

Every series must do one of these: converge absolutely, converge conditionally, or diverge.

_____ **4.** The series $\displaystyle\sum_{n=1}^{\infty} 2^{-n}n!$ _____.

    a) diverges

    b) converges absolutely

    c) converges conditionally

    d) converges, but not absolutely and not conditionally

_____ **5.** The series $\displaystyle\sum_{n=1}^{\infty} \frac{(-5)^{n+1}}{n^n}$ _____.

    a) diverges

    b) converges absolutely

    c) converges conditionally

    d) converges, but not absolutely and not conditionally

_____ 6. True, False:

If a series converges absolutely, then all rearrangements of the terms of the series will also converge.

_____ **1.** True, False:

$$\sum_{n=2}^{\infty} \frac{1}{(\ln n)^n} \text{ converges.}$$

_____ **2.** True, False:

$$\sum_{n=1}^{\infty} \frac{6}{7n + 8} \text{ converges.}$$

_____ **3.** True, False:

$$\sum_{n=1}^{\infty} \frac{(-1)^n \sqrt{n}}{n + 3} \text{ converges.}$$

_____ **4.** True, False:

$$\sum_{n=1}^{\infty} \frac{e^n}{n!} \text{ converges.}$$

Section 12.8

_____ **1.** Sometimes, Always, Never:

The interval of convergence of a power series $\sum\limits_{n=0}^{\infty} a_n(x-c)^n$ is an open interval $(c-R, c+R)$. (When $R=0$ we mean $\{0\}$ and when $R=\infty$ we mean $(-\infty, \infty)$.)

_____ **2.** True, False:

If a number p is in the interval of convergence of $\sum\limits_{n=0}^{\infty} a_n x^n$ then so is the number $\frac{p}{2}$.

_____ **3.** For $f(x) = \sum\limits_{n=0}^{\infty} \frac{(x-1)^n}{3^n}$, $f(3) =$

a) 0   b) 2   c) 3   d) $f(3)$ does not exist

_____ **4.** The interval of convergence of $\sum\limits_{n=1}^{\infty} \frac{x^n}{\sqrt{n}}$ is:

a) $[-1, 1]$ b) $[-1, 1)$ c) $(-1, 1]$ d) $(-1, 1)$

_____ **5.** The radius of convergence of $\sum\limits_{n=0}^{\infty} \frac{n(x-5)^n}{3^n}$ is:

a) $\frac{1}{3}$ b) 1   c) 3   d) ∞

Section 12.9

_____ **1.** Given that $e^x = \sum\limits_{n=0}^{\infty} \dfrac{x^n}{n!}$, a power series for xe^{x^2} is:

     a) $\sum\limits_{n=0}^{\infty} \dfrac{x^2}{n!}$ b) $\sum\limits_{n=0}^{\infty} \dfrac{x^{2n}}{n!}$

     c) $\sum\limits_{n=0}^{\infty} \dfrac{x^{2n+1}}{n!}$ d) $\sum\limits_{n=0}^{\infty} \dfrac{x^{2n}}{(n+1)!}$

_____ **2.** For $f(x) = \sum\limits_{n=0}^{\infty} \dfrac{x^{2n}}{n!}$, $f'(x) =$

     a) $\sum\limits_{n=1}^{\infty} \dfrac{x^{2n-1}}{n!}$ b) $\sum\limits_{n=1}^{\infty} \dfrac{2x^{2n-1}}{(n-1)!}$

     c) $\sum\limits_{n=1}^{\infty} \dfrac{2^n x^{2n-1}}{n!}$ d) $\sum\limits_{n=1}^{\infty} \dfrac{(2n-1)x^{2n-1}}{(n-1)!}$

_____ **3.** Using $\dfrac{1}{1-x} = \sum\limits_{n=0}^{\infty} x^n$, $\displaystyle\int \dfrac{x}{1-x^2}\,dx =$

     a) $\sum\limits_{n=0}^{\infty} \dfrac{x^{2n}}{2n}$ b) $\sum\limits_{n=0}^{\infty} \dfrac{x^{2n+1}}{2n+1}$

     c) $\sum\limits_{n=0}^{\infty} \dfrac{x^{2n+2}}{2n+2}$ d) $\sum\limits_{n=0}^{\infty} x^{2n+1}$

Section 12.10

_____ **1.** Given the Taylor Series $e^x = \sum_{n=0}^{\infty} \frac{x^n}{n!}$ a Taylor series for $e^{x/2}$ is:

a) $\sum_{n=0}^{\infty} \frac{2^n x^n}{n!}$

b) $\sum_{n=0}^{\infty} \frac{2x^n}{n!}$

c) $\sum_{n=0}^{\infty} \frac{x^n}{2^n n!}$

d) $\sum_{n=0}^{\infty} \frac{x^n}{2n!}$

_____ **2.** The nth term in the Taylor series about $x = 1$ for $f(x) = x^{-2}$ is:

a) $(-1)^{n+1}(n+1)!(x-1)^n$

b) $(-1)^n n!(x-1)^n$

c) $(-1)^{n+1}(n+1)(x-1)^n$

d) $(-1)^n n(x-1)^n$

_____ **3.** The Taylor polynomial of degree 3 for $f(x) = x(\ln x - 1)$ about $a = 1$ is $T_3(x) =$

a) $-1 + \frac{(x-1)^2}{2} - \frac{(x-1)^3}{6}$

b) $-2 + x + \frac{(x-1)^2}{2} - \frac{(x-1)^3}{6}$

c) $-1 + x - x^2 + x^3$

d) $-1 - x + x^2 - x^3$

_____ **4.** True, False:

If $T_n(x)$ is the nth Taylor polynomial for $f(x)$ at $x = c$, then $T_n^{(k)}(c) = f^{(k)}(c)$ for $k = 0, 1, \ldots, n$.

_____ **5.** If we know that $\left| f^{(n+1)}(x) \right| \leq M$ for all $|x - a| \leq d$, then the absolute value of the nth remainder of the Taylor series for $y = f(x)$ about a is less than or equal to

a) $\frac{M}{n!} |x - a|^n$

b) $\frac{M}{(n+1)!} |x - a|^{n+1}$

c) $\frac{M}{n!} |x - a|^{n+1}$

d) $\frac{M}{(n+1)!} |x - a|^n$

Section 12.11

—— **1.** $\dbinom{\frac{1}{2}}{3} =$

    a) $\dfrac{5}{16}$ b) $\dfrac{1}{16}$ c) $\dfrac{5}{8}$ d) $-\dfrac{5}{8}$

—— **2.** $\displaystyle\sum_{n=0}^{\infty} \dbinom{\frac{2}{3}}{n} x^n$ is the binomial series for:

    a) $(1 + x)^{2/3}$ b) $(1 + x)^{-2/3}$

    c) $(1 + x)^{3/2}$ d) $(1 + x)^{-3/2}$

—— **3.** Using a binomial series, the Maclaurin series for $\dfrac{1}{1 + x^2}$ is:

    a) $\displaystyle\sum_{n=0}^{\infty} \dbinom{1}{n} x^n$ b) $\displaystyle\sum_{n=0}^{\infty} \dbinom{-1}{n} x^n$

    c) $\displaystyle\sum_{n=0}^{\infty} \dbinom{1}{n} x^{2n}$ d) $\displaystyle\sum_{n=0}^{\infty} \dbinom{-1}{n} x^{2n}$

Section 12.12

_____ **1.** The quadratic approximation for $f(x) = x^6$ at $a = 1$ is:

a) $P(x) = 6(x - 1) + 30(x - 1)^2$

b) $P(x) = 1 + 6(x - 1) + 30(x - 1)^2$

c) $P(x) = 6(x - 1) + 15(x - 1)^2$

d) $P(x) = 1 + 6(x - 1) + 15(x - 1)^2$

_____ **2.** If the Maclaurin polynomial of degree 2 for $f(x) = e^x$ is used to approximate $e^{0.2}$ then the best estimate for the error with $0 < x < 1$ is:

a) $\frac{1}{24}e$ b) $\frac{1}{6}e$ c) $\frac{1}{2}e$ d) e

_____ **3.** What degree Taylor polynomial about $a = 1$ is needed to approximate $e^{1.05}$ accurate to within 0.0001?

a) $n = 2$ b) $n = 3$ c) $n = 4$ d) $n = 5$

Section 13.1

_____ **1.** The point plotted at the right has coordinates:

   a) $(2, 5, 4)$ b) $(5, 2, 4)$
   c) $(4, 2, 5)$ d) $(2, 4, 5)$

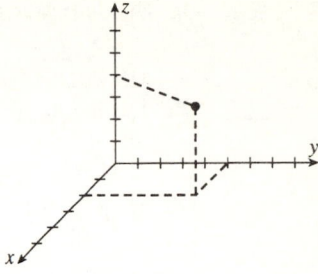

_____ **2.** The equation of the plane partially drawn is:

   a) $x = 3$
   b) $y = 3$
   c) $z = 3$
   d) $x + y = 3$

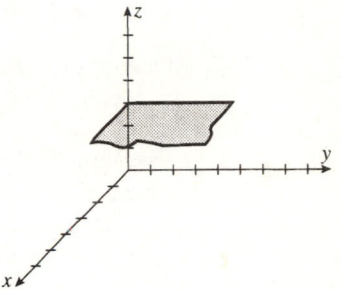

_____ **3.** The distance between $(7, 4, -3)$ and $(-1, 2, 3)$ is:

   a) 4   b) 16   c) 104   d) $\sqrt{104}$

_____ **4.** The sphere graphed at the right has equation

   a) $(x - 2)^2 + (y - 3)^2 + (z - 4)^2 = 1$
   b) $(x - 2)^2 + (y - 4)^2 + (z - 4)^2 = 4$
   c) $(x - 2)^2 + (y - 5)^2 + (z - 4)^2 = 4$
   d) $(x - 2)^2 + (y - 4)^2 + (z - 4)^2 = 1$

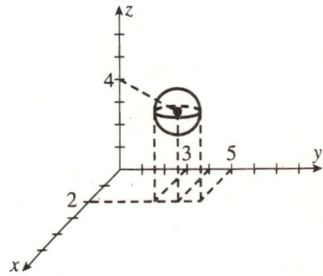

_____ **5.** The region described by $y^2 + z^2 = 4$ is:

   a) a circle, center at the origin   b) a sphere, center at the origin
   c) a cone, along the x-axis   d) a cylinder, along the x-axis

Section 13.2

___ 1. The vector represented by $\overrightarrow{AB}$ where $A: (4, 8)$, $B(6, 6)$ is:

a) $\langle -2, 2 \rangle$ b) $\langle 2, -2 \rangle$ c) $\langle 10, 14 \rangle$ d) $\langle 14, 10 \rangle$

___ 2. The length of $\mathbf{a} = 4\mathbf{i} - \mathbf{j} - 2\mathbf{k}$ is:

a) $\sqrt{11}$ b) 11 c) $\sqrt{21}$ d) 21

___ 3. For $\mathbf{a} = 6\mathbf{i} - \mathbf{j}$ and $\mathbf{b} = 2\mathbf{i} + 3\mathbf{j}$, $\mathbf{a} + 2\mathbf{b} =$

a) $2\mathbf{i} - 4\mathbf{j}$ b) $8\mathbf{i} - 2\mathbf{j}$ c) $16\mathbf{i} - 4\mathbf{j}$ d) $10\mathbf{i} + 5\mathbf{j}$

___ 4. True, False:

$\mathbf{i} + \mathbf{j}$ is a unit vector in the direction of $5\mathbf{i} + 5\mathbf{j}$.

___ 5. The vector $\mathbf{c}$ in the figure at the right is:

a) $\mathbf{a} - \mathbf{b}$
b) $\mathbf{b} - \mathbf{a}$
c) $\mathbf{a} + \mathbf{b}$
d) $-\mathbf{a} - \mathbf{b}$

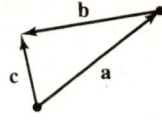

___ 6. The unit vector in the direction of $\mathbf{u} = \langle 1, 3, -1 \rangle$ is

a) $\left\langle \dfrac{1}{3}, 1, -\dfrac{1}{3} \right\rangle$

b) $\left\langle \dfrac{1}{11}, 1, -\dfrac{1}{11} \right\rangle$

c) $\left\langle \dfrac{1}{11}, \dfrac{3}{11}, -\dfrac{1}{11} \right\rangle$

d) $\left\langle \dfrac{1}{\sqrt{11}}, \dfrac{3}{\sqrt{11}}, -\dfrac{1}{\sqrt{11}} \right\rangle$

Section 13.3

_____ **1.** For $\mathbf{a} = 2\mathbf{i} + 3\mathbf{j} - 4\mathbf{k}$ and $\mathbf{b} = -\mathbf{i} + 2\mathbf{j} - \mathbf{k}$, $\mathbf{a} \cdot \mathbf{b} =$

    a) 0   b) 8   c) 10   d) 12

_____ **2.** The angle θ at the right has cosine:

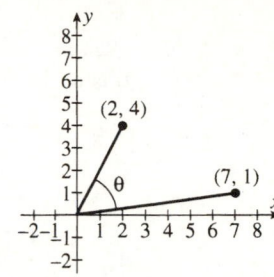

    a) $\dfrac{9}{5\sqrt{10}}$ b) $\dfrac{9}{250}$

    c) $\dfrac{9}{\sqrt{50}}$ d) $\dfrac{9}{50}$

_____ **3.** The scalar projection of $\mathbf{a} = \langle 1, 4 \rangle$ on $\mathbf{b} = \langle 6, 3 \rangle$ is:

    a) $\dfrac{18}{\sqrt{17}}$ b) $\dfrac{6}{\sqrt{5}}$ c) $\dfrac{6}{\sqrt{17}\sqrt{5}}$ d) $\dfrac{18}{\sqrt{17}\sqrt{5}}$

_____ **4.** The α, β, γ direction cosines in the figure are (respectively):

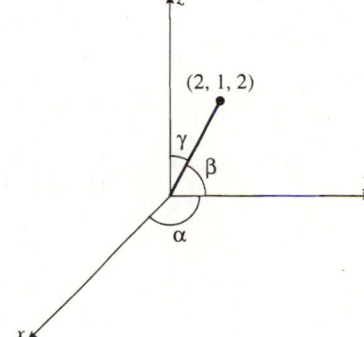

    a) $\dfrac{2}{9}, \dfrac{1}{9}, \dfrac{2}{9}$

    b) $\dfrac{2}{3}, \dfrac{1}{3}, \dfrac{2}{3}$

    c) $\sqrt{\dfrac{2}{9}}, \sqrt{\dfrac{1}{9}}, \sqrt{\dfrac{2}{9}}$

    c) $\sqrt{\dfrac{2}{3}}, \sqrt{\dfrac{1}{3}}, \sqrt{\dfrac{2}{3}}$

Section 13.4

_____ **1.** For $\mathbf{a} = \langle 1, 2, 1 \rangle$ and $\mathbf{b} = \langle 3, 0, 2 \rangle$, $\mathbf{a} \times \mathbf{b} =$

    a) $\langle -6, -1, 4 \rangle$ b) $\langle -6, 1, 4 \rangle$

    c) $\langle 4, -1, -6 \rangle$ d) $\langle 4, 1, -6 \rangle$

_____ **2.** True, False:

    $\mathbf{a} \times \mathbf{b}$ is in the plane formed by vectors $\mathbf{a}$ and $\mathbf{b}$.

_____ **3.** $\mathbf{j} \times (-\mathbf{k}) =$

    a) $\mathbf{i}$ b) $-\mathbf{i}$ c) $\mathbf{j} + \mathbf{k}$ d) $-\mathbf{j} - \mathbf{k}$

_____ **4.** The area of the parallelogram at the right is:

    a) 26

    b) $\sqrt{14}$

    c) $\sqrt{86}$

    d) $\sqrt{300}$

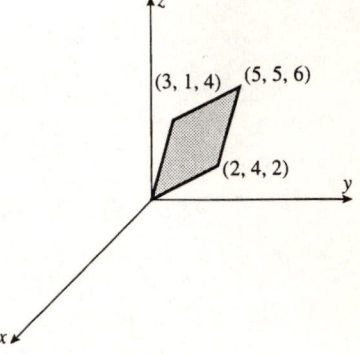

_____ **5.** For $\mathbf{a} = \langle 6, 2, 3 \rangle$, $\mathbf{b} = \langle 4, 7, 9 \rangle$, $\mathbf{c} = \langle 8, 1, 5 \rangle$, $\mathbf{c} \cdot (\mathbf{a} \times \mathbf{b}) =$

  a) $\begin{vmatrix} 6 & 2 & 3 \\ 4 & 7 & 9 \\ 8 & 1 & 5 \end{vmatrix}$ b) $\begin{vmatrix} 6 & 2 & 3 \\ 8 & 1 & 5 \\ 4 & 7 & 9 \end{vmatrix}$ c) $\begin{vmatrix} 8 & 1 & 5 \\ 6 & 2 & 3 \\ 4 & 7 & 9 \end{vmatrix}$ d) $\begin{vmatrix} 8 & 1 & 5 \\ 4 & 7 & 9 \\ 6 & 2 & 3 \end{vmatrix}$

Section 13.5

_____ **1.** The equation of the line through $(4, 10, 8)$ and $(3, 5, 1)$ is:

    a) $\dfrac{x-4}{3} = \dfrac{y-10}{5} = \dfrac{z-8}{1}$ b) $\dfrac{x-3}{4} = \dfrac{y-5}{10} = \dfrac{z-1}{8}$

    c) $\dfrac{x-3}{7} = \dfrac{y-5}{15} = \dfrac{z-1}{9}$ d) $\dfrac{x-4}{1} = \dfrac{y-10}{5} = \dfrac{z-8}{7}$

_____ **2.** True, False:

    These lines are skew: $x = 1 + 5t$ $x = 4 + 7s$

                             $y = 3 - 6t$ $y = 5 + s$

                             $z = 1 - 2t$ $z = 1 - 4s$

_____ **3.** The equation of the plane through $\langle -7, 2, 3 \rangle$ with normal vector $\langle 6, -4, 1 \rangle$ is:

    a) $6x - 4y + z = -47$ b) $6x - 4y + z = 0$

    c) $-7x + 2y + 3z = -20$ d) $-7x + 2y + 3z = -12$

_____ **4.** The equation of the plane formed by the two lines

    $x = 3 + 2t$ $x = 3 + t$

    $y = 1 - 4t$ and $y = 1 + 2t$

    $z = 5 + t$ $z = 5 + 2t$

    is:

    a) $2(x - 3) - 4(y - 1) + (z - 5) = 0$

    b) $(x - 3) + 2(y - 1) + 2(z - 5) = 0$

    c) $(x - 3) - 6(y - 1) + (z - 5) = 0$

    d) $-10(x - 3) - 3(y - 1) + 8(z - 5) = 0$

_____ **5.** The distance from $(1, 2, 1)$ to the plane $6x + 5y + 8z = 34$ is:

    a) $\sqrt{5}$ b) $2\sqrt{5}$ c) $\dfrac{2}{\sqrt{5}}$ d) $\dfrac{1}{\sqrt{5}}$

Section 13.6

_____ **1.** $\frac{x^2}{25} + 1 = \frac{z^2}{16} - \frac{y^2}{9}$ is a(n):

    a) hyperboloid of one sheet   b) hyperboloid of two sheets

    c) ellipsoid   d) hyperbolic cone

_____ **2.** $\frac{x^2}{9} + \frac{z^2}{4} = 1$ is a(n):

    a) ellipsoid   b) elliptic cone

    c) elliptic paraboloid   d) elliptic cylinder

_____ **3.** The graph at the right has equation

    a) $\frac{x^2}{a^2} + \frac{y^2}{b^2} = \frac{z^2}{c^2}$

    b) $\frac{x^2}{a^2} + \frac{z^2}{c^2} = \frac{y^2}{b^2}$

    c) $\frac{y^2}{b^2} + \frac{z^2}{c^2} = \frac{x^2}{a^2}$

    d) $\frac{x^2}{a^2} + \frac{y^2}{b^2} + \frac{z^2}{c^2} = 0$

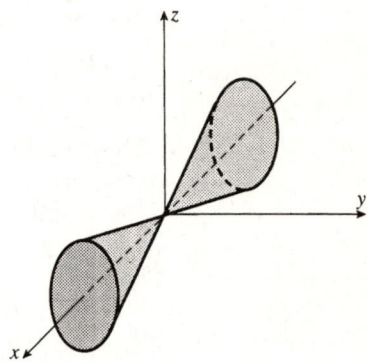

_____ **4.** The graph at the right has equation

    a) $x = \frac{y^2}{b^2} + \frac{z^2}{c^2}$

    b) $y = \frac{x^2}{a^2} + \frac{z^2}{c^2}$

    c) $z = \frac{x^2}{a^2} + \frac{y^2}{b^2}$

    d) $y^2 = \frac{x^2}{a^2} + \frac{z^2}{c^2}$

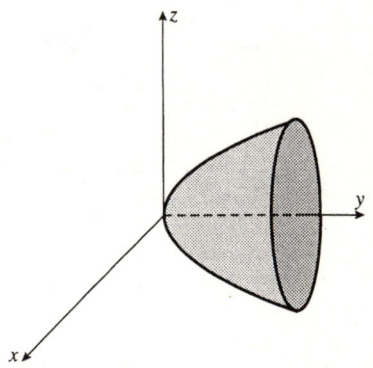

_____ **5.** $x^2 + 6x + y^2 - 10y + z = 0$ is a(n):

    a) ellipsoid

    b) paraboloid

    c) hyperbolic cone

    d) elliptic cone

_____ **6.** True, False:

The graph of $x = 5$ is a cylinder.

On Your Own

Section 13.7

_____ **1.** The spherical coordinates of the point with rectangular coordinates $(-2, 2\sqrt{3}, 4\sqrt{3})$ are:

    a) $\left(8, \frac{2\pi}{3}, \frac{\pi}{6}\right)$ b) $\left(8, \frac{\pi}{3}, -\frac{\pi}{6}\right)$ c) $\left(8, \frac{\pi}{3}, \frac{\pi}{6}\right)$ d) $\left(8, \frac{2\pi}{3}, -\frac{\pi}{6}\right)$

_____ **2.** The rectangular coordinates of the point with spherical coordinates $\left(2, \frac{\pi}{2}, \frac{\pi}{4}\right)$ are:

    a) $(2, \sqrt{2}, \sqrt{2})$ b) $(\sqrt{2}, \sqrt{2}, 0)$ c) $(\sqrt{2}, 0, \sqrt{2})$ d) $(0, \sqrt{2}, \sqrt{2})$

_____ **3.** The surface given by $z = -r^2$ is a(n)

    a) elliptic cone   b) hyperboloid of one sheet

    c) hyperbolic cylinder   d) elliptic paraboloid

_____ **4.** The point P graphed at the right has spherical coordinates

    a) $\left(2, \frac{\pi}{3}, \frac{\pi}{3}\right)$

    b) $\left(2, \frac{\pi}{3}, \frac{\pi}{6}\right)$

    c) $\left(4, \frac{\pi}{3}, \frac{\pi}{3}\right)$

    d) $\left(4, \frac{\pi}{3}, \frac{\pi}{6}\right)$

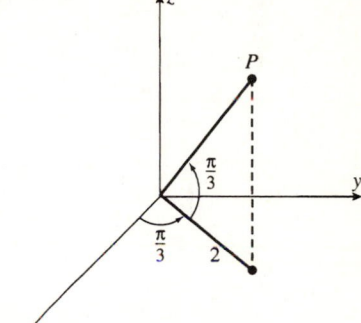

_____ **5.** The cylindrical coordinates of the point with rectangular coordinates $(5, 5, 4)$ are:

    a) $\left(5\sqrt{2}, \frac{\pi}{4}, 4\right)$ b) $\left(5\sqrt{2}, \frac{\pi}{2}, 4\right)$ c) $\left(5, \frac{\pi}{4}, 4\right)$ d) $\left(5, \frac{\pi}{2}, 4\right)$

Section 14.1

_____ **1.** The curve given by $\mathbf{r}(t) = 2\mathbf{i} + t\mathbf{j} + 2t\mathbf{k}$ is a

    a) line   b) plane   c) spiral   d) circle

_____ **2.** $\displaystyle\lim_{t \to \infty} \left\langle e^{-2t}, \cos \frac{1}{t} \right\rangle =$

    a) $\langle 1, 1 \rangle$ b) $\langle 0, 1 \rangle$ c) $\langle 0, 0 \rangle$ d) does not exist

_____ **3.** The best equation that describes the curve at the right is

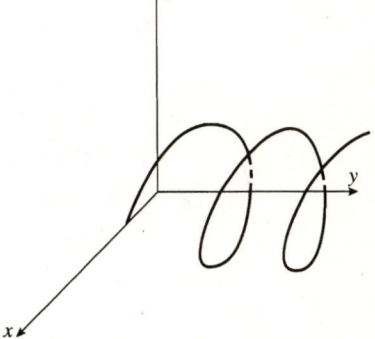

    a) $\mathbf{r}(t) = \langle t, \sin t, \cos t \rangle$
    b) $\mathbf{r}(t) = \langle \cos t, t, \sin t \rangle$
    c) $\mathbf{r}(t) = \langle \cos t, \sin t, t \rangle$
    d) $\mathbf{r}(t) = \langle \cos t, \sin t + \cos t, \sin t \rangle$

Section 14.2

_____ 1. For $\mathbf{r}(t) = t^3\mathbf{i} + \sin t\mathbf{j} - (t^2 + 2t)\mathbf{k}, \mathbf{r}'(0) =$

     a) $\mathbf{j} - 2\mathbf{k}$ b) $3\mathbf{i} - 4\mathbf{k}$ c) $3\mathbf{i} + \mathbf{j} - 2\mathbf{k}$ d) $\mathbf{0}$ (zero vector)

_____ 2. True, False:
$[\mathbf{r}(t) \times \mathbf{s}(t)]' = \mathbf{r}(t) \times \mathbf{s}'(t) + \mathbf{s}(t) \times \mathbf{r}'(t)$.

_____ 3. True, False:
$\mathbf{r}(t) = t^2\mathbf{i} - |t|\mathbf{j} + |1 - t|\mathbf{k}$, for $-2 \le t \le 2$, is piecewise smooth.

_____ 4. $\displaystyle\int_0^1 (e^t\mathbf{i} + 3\sqrt{t}\mathbf{j} + 2t\mathbf{k})dt =$

     a) $e\mathbf{i} + 2\mathbf{j} + \mathbf{k}$

     b) $e\mathbf{i} + \frac{3}{2}\mathbf{j} + 2\mathbf{k}$

     c) $(e - 1)\mathbf{i} + 2\mathbf{j} + \mathbf{k}$

     d) $(e - 1)\mathbf{i} + \frac{3}{2}\mathbf{j} + 2\mathbf{k}$

Section 14.3

____ **1.** A definite integral for the length of the curve given by $\mathbf{r}(t) = \langle t, 3 + t, t^2 \rangle$ for $1 \le t \le 2$ is:

a) $\displaystyle\int_1^2 \sqrt{2 + 4t^2}\, dt$

b) $\displaystyle\int_1^2 \sqrt{9 + 6t + 2t^2 + t^4}\, dt$

c) $\displaystyle\int_1^2 \sqrt{9 + 4t}\, dt$

b) $\displaystyle\int_1^2 2t\, dt$

____ **2.** For the vector function $\mathbf{r}(t)$ with unit tangent $\mathbf{T}$, the unit normal is:

a) $\dfrac{\mathbf{T}}{|\mathbf{T}|}$

b) $\dfrac{\mathbf{r}'}{|\mathbf{r}'|}$

c) $\dfrac{\mathbf{r}''}{|\mathbf{r}''|}$

d) $\dfrac{\mathbf{T}'}{|\mathbf{T}'|}$

____ **3.** The curvature of $\mathbf{r}(t) = -e^t\mathbf{i} + t\mathbf{j} + e^t\mathbf{k}$ at $t = 0$ is:

a) $\dfrac{\sqrt{2}}{3\sqrt{3}}$

b) $\dfrac{\sqrt{2}e}{(2e + 1)^{3/2}}$

c) $\dfrac{1}{(2e)^{3/2}}$

d) $\dfrac{\sqrt{2}e}{(2e^2 + 1)^{3/2}}$

____ **4.** True, False:

If f is twice differentiable and x_0 is an inflection point for f then the curvature of f at x_0 is 0.

____ **5.** The normal plane to $\mathbf{r}(t) = t^2\mathbf{i} - t^3\mathbf{j} + t^4\mathbf{k}$ at $t = 1$ has equation:

a) $2(x - 1) - 3(y + 1) + 4(z - 1) = 9$
b) $2(x - 1) - 3(y + 1) + 4(z - 1) = 20$
c) $(x - 2) - (y + 3) + (z - 4) = 9$
d) $(x - 2) - (y + 3) + (z - 4) = 20$

Section 14.4

_____ 1. The acceleration for a particle whose position at time t is
$3t^3\mathbf{i} + \ln t\mathbf{j} - \sin 2t\mathbf{k}$ is:

   a) $9t^2\mathbf{i} + \frac{1}{t}\mathbf{j} - 2\cos 2t\mathbf{k}$ b) $18t\mathbf{i} - \frac{1}{t^2}\mathbf{j} + 4\sin 2t\mathbf{k}$

   c) $18\mathbf{i} + \frac{2}{t^3}\mathbf{j} - 8\cos 2t\mathbf{k}$ d) $\frac{3}{4}t^4\mathbf{i} + (t\ln t - t)\mathbf{j} + \frac{1}{2}\cos 2t\mathbf{k}$

_____ 2. With initial position $\mathbf{i}$, velocity $2\mathbf{j}$, and acceleration $30t\mathbf{i} + 60t^2\mathbf{k}$, the position
function is:

   a) $5t^3\mathbf{i} + 2\mathbf{j} + 5t^4\mathbf{k}$ b) $(5t^3 + 1)\mathbf{i} + (2t + 1)\mathbf{j} + 5t^4\mathbf{k}$
   c) $(5t^3 + 1)\mathbf{i} + 2t\mathbf{j} + 5t^4\mathbf{k}$ d) $5t^3\mathbf{i} + 2\mathbf{j} + (5t^4 + 1)\mathbf{k}$

_____ 3. The force needed for a 10 kg object to attain velocity $6t^3\mathbf{i} + 10t\mathbf{j}$ is:

   a) $20t^3\mathbf{i} + 50t^2\mathbf{j}$ a) $5t^4\mathbf{i} + \frac{50}{3}t^3\mathbf{j}$

   c) $180t^2\mathbf{i} + 100\mathbf{j}$ d) $120\mathbf{i}$

_____ 4. The tangential component of acceleration for $\mathbf{r}(t) = e^t\mathbf{i} - e^{-t}\mathbf{j}$ at $t = 0$ is:

   a) 0 b) $\frac{e^4 - 1}{\sqrt{e^6 - e^2}}$ c) $\frac{e^4}{\sqrt{e^2 - 1}}$ d) $\frac{e^4 - 1}{e}$

Section 15.1

_____ **1.** Which of these points is not in the domain of $f(x, y) = \sqrt{8 - x - 2y^2}$?

a) $(0, 0)$ b) $(-6, 2)$

c) $(3, 3)$ d) All these points *are* in the domain.

_____ **2.** The level curves (for $k \neq 0$) of $f(x, y) = xy$ are:

a) ellipses   b) hyperbolas   c) parabolas   d) pairs of lines

_____ **3.** Which function best fits the graph at the right?

a) $f(x, y) = 4 - x^2$

b) $f(x, y) = 4 - y^2$

c) $f(x, y) = 4 - xy$

d) $f(x, y) = 4 - x^2 - y^2$

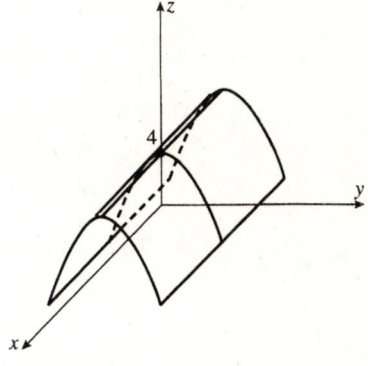

_____ **4.** The range of $f(x, y) = \sqrt{x} + \dfrac{1}{\sqrt{y}}$ is:

a) $(-\infty, \infty)$ b) $[0, \infty)$ c) $(0, \infty)$ d) $[1, \infty]$

Section 15.2

____ **1.** True, False:

If $\lim\limits_{(x, y)\to(a, b)} f(x, y) = L$ and $\lim\limits_{(x, y)\to(a, b)} g(x, y) = M \neq 0$, then
$\lim\limits_{(x, y)\to(a, b)} \dfrac{f(x, y)}{g(x, y)} = \dfrac{L}{M}$.

____ **2.** $\lim\limits_{(x, y)\to(-1, 2)} \dfrac{xy}{x^2 - y^2} =$

  a) $\dfrac{2}{3}$ b) $-\dfrac{2}{3}$ c) $-\dfrac{2}{5}$ d) does not exist

____ **3.** $\lim\limits_{(x, y)\to(0, 0)} \dfrac{\sin(xy)}{y^2} =$

  a) 0   b) 1   c) π d) does not exist

____ **4.** $f(x, y) = \sqrt{x + y}$ is continuous for all (x, y) such that:

  a) $x \geq 0$ and $y \geq 0$ b) $y > x$

  c) $y > -x$ d) $-y \leq x$

____ **5.** True, False:

$f(x, y) = \begin{cases} \dfrac{\sin x}{y} & y \neq 0 \\ 1 & y = 0 \end{cases}$ is continuous at $(0, 0)$.

Section 15.3

_____ **1.** $f_x(a, b)$ is the slope of the tangent line to the surface $z = f(x, y)$ at (a, b) determined by:

a) the trace through $x = a$
b) the trace through $y = b$
c) the trace through $x = y$
d) the intersection of the two traces $x = a$ and $y = b$

_____ **2.** For $f(x, y) = e^{x^2 y^3}$, $f_y(x, y) =$

a) $3y^2 e^{x^2 y^3}$

b) $3x^2 y^2 e^{x^2 y^3}$

c) $(2xy^3 + 3x^2 y^3)e^{x^2 y^3}$

d) $e^{3x^2 y^3}$

_____ **3.** For $f(x, y) = \ln(2x + 3y)$, $f_{yx} =$

a) $\dfrac{-6}{(2x + 3y)^2}$

b) $\dfrac{3}{2x + 3y}$

c) $\dfrac{6}{2x + 3y}$

d) $\dfrac{2}{x} + \dfrac{3}{y}$

_____ **4.** Sometimes, Always, or Never:
$f_{xy} = f_{yz}$

_____ **5.** How many second partial derivatives will there be for $z = f(r, s, t, x, y)$? (Do not assume that they are continuous.)

a) 5   b) 10   c) 20   d) 25

Section 15.4

_____ **1.** True, False:

If dz exists at (x, y) then $z = f(x, y)$ is differentiable at (x, y).

_____ **2.** The equation of the tangent plane to $z = x^2 + xy + y^3$ for $(x, y) = (2, -1)$ is:

a) $3x + 5y - z = 0$ b) $3x + 5y - z = 2$

c) $5x + 3y - z = 0$ d) $5x + 3y - z = 2$

_____ **3.** For $z = \sin(xy)$, $dz =$

a) $xy \cos(xy)dx + xy \cos(xy)dy$

b) $x \cos(xy)dx + y \cos(xy)dy$

c) $y \cos(xy)dx + x \cos(xy)dy$

d) 0

_____ **4.** The volume of a pyramid is $\frac{1}{3}$ area of its base times its altitude. Using differentials, an estimate for the volume of the pyramid at the right is:

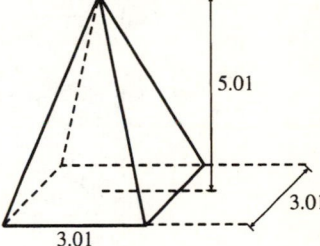

a) 15.10

b) 15.12

c) 15.13

d) 15.14

Section 15.5

____ **1.** For $z = xy^2$, $x = t + v^2$, $y = t^2 + v$, $\frac{\partial z}{\partial v} =$

    a) $2[(t^2 + v)^2 + (t^2 + v)(t + v^2)]$
    b) $2[(t^2 + v)^2 + (t + v^2)^2]$
    c) $2(t^2 + v)(2t^2v + v^2)$
    d) $2(t^2 + v)(t^2v + t + 2v^2)$

____ **2.** Find $\frac{\partial f}{\partial r}$ for $f(x, y) = x^2 + 2y^2$, $x = 2rs$, $y = 4r^2s^2$.

    a) $8rs^2 + 128r^3s^4$ b) $2r^2s^2 + 64r^3s^4$
    c) $4rs^2 + 64r^3s^4$ d) $2r^2s^2 + 128r^3s^4$

____ **3.** Find $\frac{dy}{dx}$ for y defined by $x^2 - 6xy + y^2 = 20$.

    a) $\dfrac{x - 3y}{3x - y}$ b) $\dfrac{x - 3y}{y - 3x}$ c) $\dfrac{y - 3x}{x - 3y}$ b) $\dfrac{3x - y}{x - 3y}$

____ **4.** Find $\frac{\partial z}{\partial x}$ for $z = f(x, y)$ defined implicitly by $\frac{x^2}{4} - \frac{y^2}{9} + \frac{z^2}{16} = 0$.

    a) $\dfrac{-x}{4z}$ b) $\dfrac{x}{4z}$ c) $\dfrac{-4x}{z}$ b) $\dfrac{4x}{z}$

Section 15.6

___ **1.** For $f(x, y) = x^3y + xy^2$ and $\mathbf{u} = \left\langle \frac{3}{5}, -\frac{4}{5} \right\rangle$, $D_{\mathbf{u}}f(-1, 1) =$

  a) $\dfrac{12}{5}$ b) $-\dfrac{12}{5}$ c) $\dfrac{24}{5}$ d) $-\dfrac{24}{5}$

___ **2.** True, False:
$D_{\mathbf{u}}f(a, b) = \nabla f(a, b) \cdot \mathbf{u}$.

___ **3.** In what direction $\mathbf{u}$ is $D_{\mathbf{u}}f(-1, 1)$ maximum for $f(x, y) = x^3y^4$?

  a) $\langle 3, -4 \rangle$ b) $\left\langle \frac{3}{5}, -\frac{4}{5} \right\rangle$ c) $\langle 4, -3 \rangle$ d) $\left\langle \frac{4}{5}, -\frac{3}{5} \right\rangle$

___ **4.** The equation of the plane tangent to $x^2 + 2y^2 + 2z^2 = 5$ at $(-1, -1, 1)$ is:

  a) $-(x + 1) - (y + 1) + (z - 1) = 0$
  b) $-(x - 1) - (y - 1) + (z + 1) = 0$
  c) $-2(x + 1) - 4(y + 1) + 4(z - 1) = 0$
  d) $-4(x + 1) - 4(y + 1) - 2(z - 1) = 0$

___ **5.** Which vector at the right could be a gradient for $f(x, y)$?

  a) **a**
  b) **b**
  c) **c**
  d) **d**

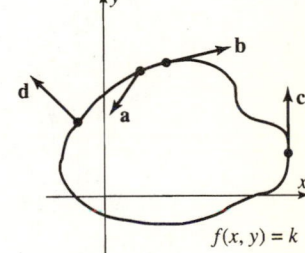

$f(x, y) = k$

Section 15.7

_____ **1.** A critical point of $f(x, y) = 60 - 6x + x^2 + 8y - y^2$ is:

   a) $(3, 4)$ b) $(-3, -4)$ c) $(-3, 4)$ d) $(3, -4)$

_____ **2.** If (a, b) is a critical point of $f(x, y)$, all second derivatives of f are continuous, and $f_{xx}(a, b) = 10, f_{yy}(a, b) = 3, f_{xy}(a, b) = 5$, then (a, b) is a:

   a) local maximum
   b) local minimum
   c) saddle point
   d) cannot tell from the information given

_____ **3.** The absolute minimum of $f(x, y) = 40 - 3x^2 - 2y^2$ with domain $D = \{(x, y) \mid -2 \le x \le 3, -1 \le y \le 2\}$ is:

   a) 0   b) 5   c) 12   d) 40

_____ **4.** A box with half a lid is to hold 48 cubic inches. If the box is to have minimum surface area, then the value of y must be:

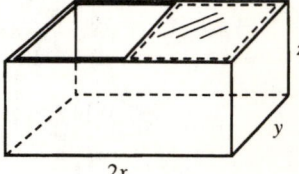

   a) 4   b) $\sqrt[3]{48}$
   c) 6   d) 8

_____ **5.** True, False:

If $f_{xx}f_{yy} - (f_{xy})^2 = 0$ at (x_0, y_0) then (x_0, y_0) cannot be a saddle point.

Section 15.8

_____ **1.** The absolute maximum of $f(x, y) = 12x - 8y$ subject to $x^2 + 2y^2 = 11$ occurs at:

a) $(3, -1)$ b) $(-3, 1)$ c) $(\sqrt{11}, 0)$ d) $\left(0, -\sqrt{\dfrac{11}{2}}\right)$

_____ **2.** To find the minimum surface area of a rectangular box whose volume is 100 cm^3 and total edge length is 50 cm by Lagrange multipliers, which set of equations results?

a) $2y + 2z = \lambda_1 yz + 4\lambda_2$
$2x + 2z = \lambda_1 xz + 4\lambda_2$
$2x + 2y = \lambda_1 xy + 4\lambda_2$
$xyz = 100$
$4x + 4y + 4z = 50$

b) $yz = \lambda_1(2y + 2z) + 4\lambda_2$
$xz = \lambda_1(2x + 2z) + 4\lambda_2$
$xy = \lambda_1(2x + 2y) + 4\lambda_2$
$xyz = 100$
$4x + 4y + 4z = 50$

c) $4y + 4z = \lambda_1 yz + \lambda_2(2y + 2z)$
$4x + 4z = \lambda_1 xz + \lambda_2(2x + 2z)$
$4x + 4y = \lambda_1 xy + \lambda_2(2x + 2z)$
$xyz = 100$
$4x + 4y + 4z = 50$

d) $4 = \lambda_1 yz + \lambda_2(2y + 2z)$
$4 = \lambda_1 xz + \lambda_2(2x + 2z)$
$4 = \lambda_1 xy + \lambda_2(2x + 2z)$
$xyz = 100$
$4x + 4y + 4z = 50$

Section 16.1

_____ **1.** How many subrectangles of $[0, 4] \times [3, 8]$ are determined by the partition with $\Delta x = 2$ and $\Delta y = 1$?

   a) 6   b) 8   c) 10   d) 20

_____ **2.** Find the Riemann sum for $\iint\limits_{R} (x + y)\,dA$ using midpoints and the partition of R given at the right.

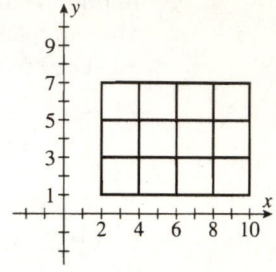

   a) 120
   b) 240
   c) 360
   d) 480

_____ **3.** Sometimes, Always, or Never:

$\iint\limits_{R} f(x, y)\,dA$ is the volume above the rectangular region R and under $z = f(x, y)$.

Section 16.2

_____ **1.** Evaluate $\int_0^2 \int_0^1 6xy^2 \, dx \, dy$.

    a) 6   b) 8   c) 24   d) 30

_____ **2.** True, False:

$$\int_1^5 \int_2^4 (x^2 + y^2) dx \, dy = \int_2^4 \int_1^5 (x^2 + y^2) dx \, dy.$$

_____ **3.** An iterated integral for the volume of the solid shown is:

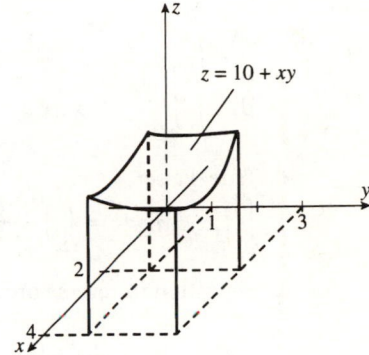

    a) $\int_2^4 \int_1^3 (10 + xy) dx \, dy$

    b) $\int_2^4 \int_1^3 (10 + xy) dy \, dx$

    c) $\int_1^2 \int_3^4 (10 + xy) dx \, dy$

    d) $\int_1^2 \int_3^4 (10 + xy) dy \, dx$

Section 16.3

____ **1.** Write $\displaystyle\iint_D x\, dA$ as an iterated integral, where D is shown at the right.

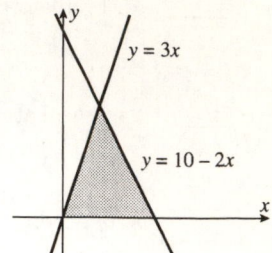

a) $\displaystyle\int_0^2 \int_{10-2x}^{3x} x\, dx\, dy$

b) $\displaystyle\int_0^5 \int_{3x}^{10-2x} x\, dy\, dx$

c) $\displaystyle\int_0^2 \int_{(x+10)/2}^{x/3} x\, dy\, dx$

d) $\displaystyle\int_0^6 \int_{y/3}^{(10-y)/2} x\, dx\, dy$

____ **2.** True, False:
$$\int_1^3 \int_2^5 xy^2\, dy\, dx = \int_2^5 \int_1^3 xy^2\, dy\, dx.$$

____ **3.** Rewritten in reverse order, $\displaystyle\int_0^1 \int_{y-1}^0 x^2 y^2\, dx\, dy =$

a) $\displaystyle\int_{-1}^1 \int_0^{x-1} x^2 y^2\, dy\, dx$

b) $\displaystyle\int_{-1}^0 \int_0^{x+1} x^2 y^2\, dy\, dx$

c) $\displaystyle\int_0^1 \int_0^{x+1} x^2 y^2\, dy\, dx$

d) $\displaystyle\int_0^1 \int_0^{x-1} x^2 y^2\, dy\, dx$

____ **4.** The area of the region at the right is:

a) $\displaystyle\int_0^2 \int_0^4 dy\, dx$

b) $\displaystyle\int_0^2 \int_0^{4-2x} dy\, dx$

c) $\displaystyle\int_0^4 \int_0^{(y-4)/2} dx\, dy$

d) $\displaystyle\int_0^4 \int_0^2 dx\, dy$

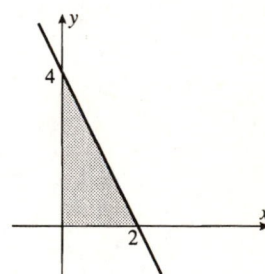

Section 16.4

____ **1.** Rewritten as an iterated integral in polar coordinates $\int_0^2 \int_0^{\sqrt{4-x^2}} y\, dy\, dx =$

   a) $\int_0^2 \int_0^{\pi/2} r \sin\theta\, d\theta\, dr$

   b) $\int_0^2 \int_0^{\pi/2} r^2 \sin\theta\, d\theta\, dr$

   c) $\int_0^{\pi/2} \int_0^2 r^2 \cos\theta\, dr\, d\theta$

   d) $\int_0^{\pi/2} \int_0^2 r \cos\theta\, dr\, d\theta$

____ **2.** An iterated integral for the area of the shaded region D is:

   a) $\int_0^5 \int_0^r 5 \cos 2\theta\, d\theta\, dr$

   b) $\int_0^5 \int_0^r r\, d\theta\, dr$

   c) $\int_0^{\pi/4} \int_0^{5\cos 2\theta} dr\, d\theta$

   d) $\int_0^{\pi/4} \int_0^{5\cos 2\theta} r\, dr\, d\theta$

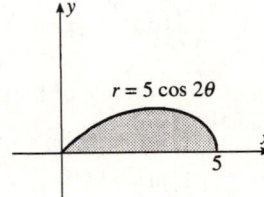

$r = 5 \cos 2\theta$

Section 16.5

_____ **1.** The mass of the lamina at the right which has a density of $x^2 + y^2$ at each point (x, y) is given by:

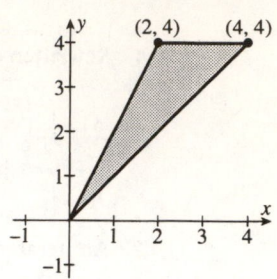

a) $\displaystyle\int_0^4 \int_{y/2}^y (x^2 + y^2)\,dx\,dy$

b) $\displaystyle\int_0^4 \int_x^{2x} (x^2 + y^2)\,dy\,dx$

c) $\displaystyle\int_0^4 \int_y^{2y} (x^2 + y^2)\,dx\,dy$

d) $\displaystyle\int_2^4 \int_{y/2}^y (x^2 + y^2)\,dx\,dy$

_____ **2.** The y coordinate of the center of mass of a lamina D with density $\rho(x, y)$ and mass m is:

a) $\displaystyle\iint_D y\rho(x, y)\,dA$

b) $\displaystyle\iint_D x\rho(x, y)\,dA$

c) $\dfrac{\displaystyle\iint_D y\rho(x, y)\,dA}{m}$

d) $\dfrac{\displaystyle\iint_D x\rho(x, y)\,dA}{m}$

_____ **3.** Let $f(x, y) = \begin{cases} x + y^2 & \text{if } 0 \le x \le 1, 0 \le y \le x \\ 0 & \text{otherwise} \end{cases}$

Is $f(x, y)$ a joint probability function?

Section 16.6

---

1. Write an iterated integral for the area of the plane $z = x$ that is above the region D below.

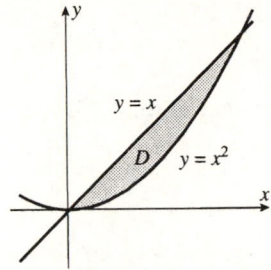

a) $\displaystyle \int_0^1 \int_x^{x^2} \sqrt{x^2 + 1}\, dy\, dx$

b) $\displaystyle \int_0^1 \int_{x^2}^x \sqrt{x^2 + 1}\, dy\, dx$

c) $\displaystyle \int_0^1 \int_{x^2}^x \sqrt{2}\, dy\, dx$

d) $\displaystyle \int_0^1 \int_y^{\sqrt{y}} \sqrt{y + 1}\, dx\, dy$

Section 16.7

_____ **1.** True, False:

$$\int_3^6 \int_1^5 \int_2^4 xyz \, dx \, dz \, dy = \int_1^4 \int_3^6 \int_2^5 xyz \, dy \, dx \, dz.$$

_____ **2.** $\displaystyle\int_1^2 \int_0^x \int_y^{x+y} 12x \, dz \, dy \, dx =$

a) 45   b) 15   c) 63   d) 127

_____ **3.** $\displaystyle\iiint_E z \, dV$, where E is the wedge-shaped solid shown at the right, equals:

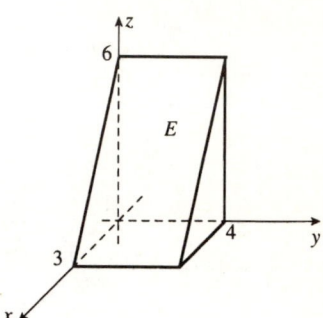

a) $\displaystyle\int_0^4 \int_0^3 \int_0^{6-2x} z \, dz \, dx \, dy$

b) $\displaystyle\int_0^4 \int_0^3 \int_0^6 z \, dz \, dx \, dy$

c) $\displaystyle\int_0^4 \int_0^3 \int_0^{6-z} z \, dx \, dz \, dy$

d) $\displaystyle\int_0^4 \int_0^3 \int_0^{6-x-y} z \, dz \, dx \, dy$

_____ **4.** The moment about the xz plane of a solid E whose density is $\rho(x, y, z) = x$ is:

a) $\displaystyle\iiint_E x^2 z \, dV$ b) $\displaystyle\iiint_E xy \, dV$

c) $\displaystyle\iiint_E x \, dV$ d) $\displaystyle\iiint_E xyz \, dV$

Section 16.8

1. Write $\iiint_E (x^2 + y^2 + z^2)\,dV$ in cylindrical coordinates, where E is the solid at the right.

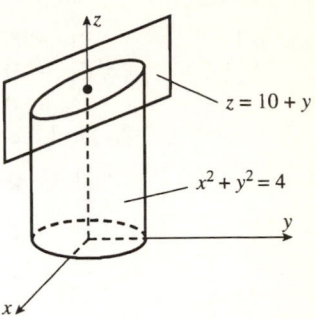

a) $\int_0^{2\pi} \int_0^2 \int_0^{10\,+y} (r^2 + z^2)r\,dz\,dr\,d\theta$

b) $\int_0^{2\pi} \int_0^4 \int_0^{x^2\,+y^2} (r^2 + z^2)r\,dz\,dr\,d\theta$

c) $\int_0^{2\pi} \int_0^2 \int_0^{10\,+y} (r^2 + z^2)\,dz\,dr\,d\theta$

d) $\int_0^{2\pi} \int_0^4 \int_0^{x^2\,+y^2} (r^2 + z^2)\,dz\,dr\,d\theta$

2. Write $\iiint_E 2\,dV$ in spherical coordinates, where E is the bottom half of the sphere at the right.

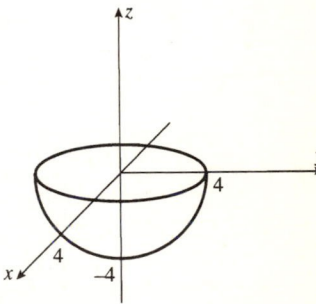

a) $\int_{\pi/2}^{\pi} \int_0^{2\pi} \int_0^2 2\rho^2 \sin\phi\,d\rho\,d\theta\,d\phi$

b) $\int_{\pi/2}^{\pi} \int_0^{2\pi} \int_0^4 2\rho^2 \sin\phi\,d\rho\,d\theta\,d\phi$

c) $\int_{\pi/2}^{\pi} \int_0^{2\pi} \int_0^2 2\,d\rho\,d\theta\,d\phi$

d) $\int_{\pi/2}^{\pi} \int_0^{2\pi} \int_0^4 2\,d\rho\,d\theta\,d\phi$

Section 16.9

_____ **1.** Find the Jacobian for $x = u^2v^2$, $y = u^2 + v^2$.

    a) $2uv^3 - 2u^3v$ b) $4u^2v - 4uv^2$

    c) $2u^2v - 2uv^2$ d) $4uv^3 - 4u^3v$

_____ **2.** Find the iterated integral for $\iint\limits_R dA$, where R is the first quadrant of the ellipse $4x^2 + 9y^2 = 36$ and the transformation is $x = 3u \cos v$, $y = 2u \sin v$.

    a) $\displaystyle\int_0^1 \int_0^{\pi/2} du\, dv$ b) $\displaystyle\int_0^1 \int_0^{\pi/2} 6u\, du\, dv$

    c) $\displaystyle\int_0^1 \int_0^{\pi/2} 3u^2\, du\, dv$ d) $\displaystyle\int_0^1 \int_0^{\pi/2} u\, du\, dv$

Section 17.1

1. For $\mathbf{F}(x, y) = (x + y)\mathbf{i} + (2x - 4y)\mathbf{j}$ which vector, **a**, **b**, **c**, or **d**, best represents $\mathbf{F}(1, 1)$?

   a) **a**

   b) **b**

   c) **c**

   d) **d**

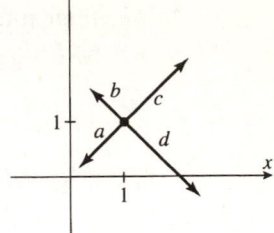

2. The gradient vector field for $f(x, y) = x^2y^3$ is:

   a) $x^2\mathbf{i} + y^3\mathbf{j}$ b) $2x\mathbf{i} + 3y^2\mathbf{j}$

   c) $2xy^3\mathbf{i} + 3x^2y^2\mathbf{j}$ d) $y^3\mathbf{i} + x^2\mathbf{j}$

3. If $\nabla f = \mathbf{F}$ the vector field $\mathbf{F}$ is:

   a) linear   b) conservative

   c) gradient   d) tangent

Section 17.2

____ **1.** A definite integral for $\int_C (x+y)ds$, where C is the curve given by $x = 3t$, $y = t, t \in [0, 1]$ is:

   a) $\int_0^1 4t\, dt$ b) $\int_0^1 4\sqrt{10}\, t\, dt$

   c) $\int_0^1 4t\sqrt{10t}\, dt$ d) $\int_0^1 4t\sqrt{t}\, dt$

____ **2.** True, False:
   The curve below appears to be piecewise smooth.

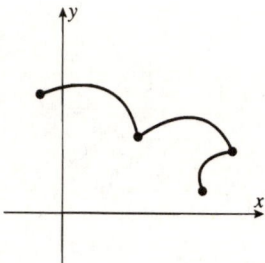

____ **3.** A definite integral for $\int_C \mathbf{F}\, d\mathbf{r}$ where $\mathbf{F}(x, y) = x^2\mathbf{i} + y^2\mathbf{j}$ and C is the line from $(1, 0)$ to $(0, 1)$ given by $x = 1 - t, y = t, t \in [0, 1]$ is:

   a) $\int_0^1 (2t - 1)dt$ b) $\int_0^1 t^2 + (1 - t)^2\, dt$

   c) $\int_0^1 t^3 + (1 - t)^2 t\, dt$ d) $\int_0^1 2\, dt$

____ **4.** True, False:
   $$\int_C f(x, y)dy = \int_a^b f(x(t), y(t))dt.$$

____ **5.** The mass of a thin wire shaped like the curve $y = x^2$ from $(0, 0)$ to $(1, 1)$ with density $1 + x^2 + y^2$ at (x, y) is

   a) $\int_0^1 (1 + t^2 + t^4)\sqrt{1 + 4t^2}\, dt$ b) $\int_0^1 (1 + t^2 + t^4)dt$

   c) $\int_0^1 \sqrt{1 + 4t^2}\, dt$ d) $\int_0^1 \sqrt{5}(1 + t^2 + t^4)dt$

Section 17.3

____ **1.** For $f(x, y) = x^2 - y^2$ and C, the parabola $y = x^2$ from $(0, 0)$ to $(2, 4)$,

$$\int_C \nabla f \cdot d\mathbf{r} =$$

    a) 0   b) 2   c) -2 d) -12

____ **2.** Which curve is simple but not closed?

    a) 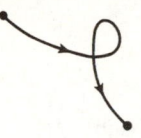   b)

    c)    d)

____ **3.** True, False:

    $\mathbf{F}(x, y) = (x^2 + 2y^2)\mathbf{i} + (y^2 + 2x^2)\mathbf{j}$ is conservative.

____ **4.** Let C be the line from $(1, 0)$ to $(0, 1)$ and $\mathbf{F}(x, y) = y\mathbf{i} + (x + y)\mathbf{j}$. $\int_C \mathbf{F} \cdot d\mathbf{r} =$

    a) $\dfrac{1}{2}$ b) 1   c) 2   d) 4

Section 17.4

____ **1.** True, False:

If $\nabla f = \mathbf{F} = P\mathbf{i} + Q\mathbf{j}$, then $\oint_C \mathbf{F} \cdot d\mathbf{r} = \iint_D f(x, y)\, dA$.

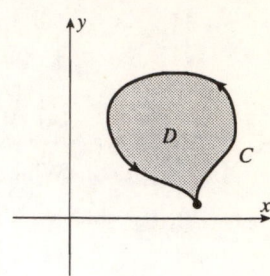

____ **2.** Let C be the curve that bounds the rectangle at the right. For $\mathbf{F}(x, y) = (x^3 + y)\mathbf{i} + (2xy)\mathbf{j}$, write a double integral for $\oint_C \mathbf{F} \cdot d\mathbf{r}$.

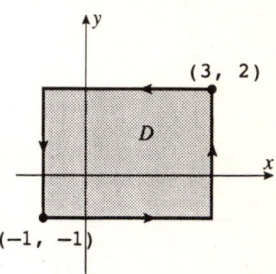

a) $\displaystyle\iint_D (2y - 1)\, dA$

b) $\displaystyle\iint_D (2xy - x^2 y)\, dA$

c) $\displaystyle\iint_D (2x - 3x^2)\, dA$

d) $\displaystyle\iint_D (x^3 + y - 2xy)\, dA$

____ **3.** The area of a region D enclosed by curve C, such as that pictured in question 1 above, may be described as:

a) $\dfrac{1}{2}\oint_C x\, dy$

b) $\dfrac{1}{2}\oint_C y\, dx$

c) $\dfrac{1}{2}\oint_C x\, dy + y\, dx$

d) $\dfrac{1}{2}\oint_C x\, dy - y\, dx$

Section 17.5

_____ **1.** Find the curl of $\mathbf{F}(x, y, z) = (x^2 + y^2)\mathbf{i} + (xz)\mathbf{j} + (yz)\mathbf{k}$.

    a) $-z\mathbf{i} + 2y\mathbf{j} + x\mathbf{k}$ b) $(z - x)\mathbf{i} + (z - 2y)\mathbf{k}$

    c) $(x - z)\mathbf{i} + 2x\mathbf{j} + 2y\mathbf{k}$ d) $y\mathbf{k}$

_____ **2.** True, False:
If $\mathbf{F}$ is conservative then curl $\mathbf{F} = \mathbf{0}$.

_____ **3.** Find div $\mathbf{F}$ for $\mathbf{F}(x, y, z) = (x^2 + y^2)\mathbf{i} + (xz)\mathbf{j} + (yz)\mathbf{k}$.

    a) $2z - x - y$ b) $2x + 3y + 2z$

    c) $2x + y$ d) $y + x$

_____ **4.** True, False:
div curl $\mathbf{F} = 0$.

_____ **5.** The vector form of Green's Theorem says that $\oint_C \mathbf{F} \cdot d\mathbf{r} =$

    a) $\iint_D \text{curl } \mathbf{F} \, dA$ b) $\iint_D \text{div } \mathbf{F} \, dA$

    c) $\iint_D (\text{curl } \mathbf{F}) \cdot \mathbf{k} \, dA$ d) $\iint_D (\text{div } \mathbf{F}) \cdot \mathbf{k} \, dA$

Section 17.6

_____ **1.** A parametrization of the cylinder in the figure is:

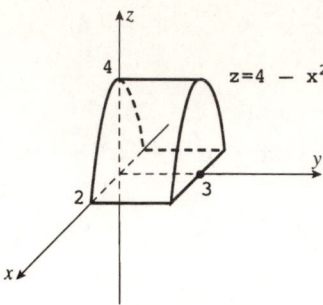

a) $x = u, y = v, z = v - u^2, -2 \le u \le 2, 0 \le v \le 3$
b) $x = u, y = 3, z = 4 - v^2, -2 \le u \le 2, 0 \le v \le z$
c) $x = u, y = v, z = 4 - u^2, -2 \le u \le 2, 0 \le v \le 3$
d) $x = u, z = 4 - v^2, -2 \le u \le 2, 0 \le v \le 2$

_____ **2.** Find the normal vector to the tangent plane at the point corresponding to $(u, v) = (1, 2)$ on the surface given by $x = 2u, y = u^2 + v^2, z = 3v$.

a) $6\mathbf{i} + 6\mathbf{j} - 8\mathbf{k}$ b) $-6\mathbf{i} - 6\mathbf{j} + 8\mathbf{k}$
c) $6\mathbf{i} - 6\mathbf{j} + 8\mathbf{k}$ d) $-6\mathbf{i} + 6\mathbf{j} - 8\mathbf{k}$

_____ **3.** A surface S is parameterized by $x = \sin u, y = \sin v, z = \cos v$ for $(u, v) \in D$, $0 \le u \le \frac{\pi}{2}, 0 \le v \le \frac{\pi}{2}$. A double integral for the surface area of S is:

a) $\displaystyle\iint_D \sqrt{(\cos u \cos v)^2 + (\sin u \sin v)^2}\, dA$

b) $\displaystyle\iint_D \sqrt{(\cos u \sin v)^2 + (\cos v \sin u)^2}\, dA$

c) $\displaystyle\iint_D \cos u\, dA$

d) $\displaystyle\iint_D \sqrt{3}\, dA$

Section 17.7

_____ 1. Find an iterated integral for the surface integral of $f(x, y, z) = x + y$ over the surface S given by

$$\mathbf{r}(u, v) = (u + v)\mathbf{i} + (u - v)\mathbf{j} + (uv)\mathbf{k}, 0 \le u \le 1, 0 \le v \le 1.$$

  a) $\displaystyle\int_0^1 \int_0^1 2u \sqrt{2u^2 + 2v^2} \, du \, dv$

  b) $\displaystyle\int_0^1 \int_0^1 \sqrt{2u^2 + 2v^2} \, du \, dv$

  c) $\displaystyle\int_0^1 \int_0^1 2u \sqrt{2u^2 + 2v^2 + 4} \, du \, dv$

  d) $\displaystyle\int_0^1 \int_0^1 \sqrt{2u^2 + 2v^2 + 4} \, du \, dv$

_____ 2. Find an iterated integral for $\displaystyle\iint_S \mathbf{F} \cdot d\mathbf{S}$ where S is the surface

$$\mathbf{r}(u, v) = u^2\mathbf{i} + v^2\mathbf{j} + uv\mathbf{k}, 0 \le u \le 1, 0 \le v \le 1, \text{ and } \mathbf{F}(x, y, z) = x\mathbf{i} + y\mathbf{j} + 2z\mathbf{k}.$$

  a) $\displaystyle\int_0^1 \int_0^1 2uv \, du \, dv$

  b) $\displaystyle\int_0^1 \int_0^1 4u^2v^2 \, du \, dv$

  c) $\displaystyle\int_0^1 \int_0^1 (2u^2 + 2v^2) \, du \, dv$

  d) $\displaystyle\int_0^1 \int_0^1 (2u^2 + 2v^2 + 4u^2v^2) \, du \, dv$

Section 17.8

_____ 1. The surface S has as its boundary the simple closed curve C. By Stokes'
Theorem $\iint\limits_{S}$ curl $\mathbf{F} \cdot d\mathbf{S} =$

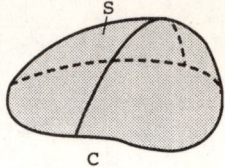

a) $\int_{C} \mathbf{F} \cdot d\mathbf{S}$

b) $\int_{C} \mathbf{F} \cdot d\mathbf{r}$

c) $\int_{C}$ div $\mathbf{F} \cdot d\mathbf{S}$

d) $\int_{C}$ div $\mathbf{F} \cdot d\mathbf{r}$

_____ 2. Use Stokes' Theorem to rewrite $\int_{C} \mathbf{F} \cdot d\mathbf{r}$, where $\mathbf{F}(x, y, z) = z\mathbf{i} - x\mathbf{j} + y\mathbf{k}$
and C is the circle $x^2 + y^2 = 36, z = 0$. Use the surface $S: z = 36 - x^2 - y^2, z \geq 0$.

a) $\int_{-6}^{6} \int_{-\sqrt{36 - u^2}}^{\sqrt{36 - u^2}} (2u + 2v - 1)dv \, du$

b) $\int_{-6}^{6} \int_{-\sqrt{36 - u^2}}^{\sqrt{36 - u^2}} (2u - 2v - 1)dv \, du$

c) $\int_{-6}^{6} \int_{-\sqrt{36 - u^2}}^{\sqrt{36 - u^2}} (2u + 2v + 1)dv \, du$

d) $\int_{-6}^{6} \int_{-\sqrt{36 - u^2}}^{\sqrt{36 - u^2}} (2u - 2v + 1)dv \, du$

Section 17.9

_____ **1.** Let S be the surface of the closed cylindrical "half can" below and
$\mathbf{F}(x, y, z) = 2xy^2\mathbf{i} + xz^2\mathbf{j} + yz\mathbf{k}$. Write a triple iterated integral for $\iint\limits_{S} \mathbf{F} \cdot d\mathbf{S}$.

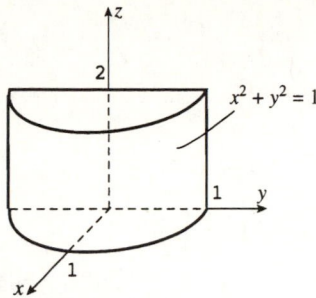

a) $\displaystyle\int_{0}^{2} \int_{0}^{1} \int_{-\sqrt{1-x^2}}^{\sqrt{1-x^2}} (4xy + 2xz + y)\,dy\,dx\,dz$

b) $\displaystyle\int_{0}^{2} \int_{0}^{1} \int_{-\sqrt{1-x^2}}^{\sqrt{1-x^2}} (2y^2 - z)\,dy\,dx\,dz$

c) $\displaystyle\int_{0}^{2} \int_{-1}^{1} \int_{0}^{\sqrt{1-x^2}} (4xy + 2xz + y)\,dx\,dy\,dz$

d) $\displaystyle\int_{0}^{2} \int_{-1}^{1} \int_{0}^{\sqrt{1-y^2}} (2y^2 + y)\,dx\,dy\,dz$

Section 18.1

_____ **1.** True, False:

If y_1 and y_2 are linearly independent solutions to
$P(x)y'' + Q(x)y' + R(x)y = 0$, then all solutions have the form $c_1y_1 + c_2y_2$.

_____ **2.** The general solution to $y'' - 6y' + 8y = 0$ is:

a) $y = e^x(c_1 \cos 4x + c_2 \sin 4x)$
b) $y = e^{4x}(c_1 \cos x + c_2 \sin x)$
c) $y = c_1e^{2x} + c_2e^{4x}$
d) $y = c_1e^{2x} + c_2xe^{4x}$

_____ **3.** The solution to $y'' - 10y' + 25y = 0, y(0) = 5, y(1) = 0$ is:

a) $y = 5e^{5x} - xe^{5x}$ b) $y = 5e^{5x} - 5xe^{5x}$
c) $y = e^{5x} - 5xe^{5x}$ d) $y = 5e^{5x} + xe^{5x}$

_____ **4.** True, False

An initial-value problem specifies $y(x_0) = y_0$ and $y(x_1) = y_1$ in the statement of the problem.

_____ 1. If $y_c(x)$ is the general solution to $ay'' + by' + cy = 0$ and $y_p(x)$ is a particular solution to $ay'' + by' + cy = G(x)$, then the general solution to the nonhomogeneous equation is $y(x) =$

a) $y_c(x) + y_p(x)$ b) $y_c(x) - y_p(x)$

c) $y_p(x) - y_c(x)$ d) $y_c(x)y_p(x)$

_____ 2. By the method of undetermined coefficients, a particular solution to $y'' - 5y' + 6y = e^{4x}$ will have the form:

a) $y_p(x) = Ae^x$ b) $y_p(x) = A_1e^x + A_2e^{6x}$

c) $y_p(x) = A_1e^{2x} + A_2e^{3x}$ d) $y_p(x) = Ae^{4x}$

_____ 3. By the method of variation of parameters, a particular solution to $y'' - 2y' + y = 3e^{4x}$ is $u_1y_1 + u_2y_2$, where $y_1 = e^x$, $y_2 = xe^x$, and

a) $u_1 = -xe^{3x} + e^{3x}, u_2 = -e^{3x}$

b) $u_1 = -xe^{3x}, u_2 = e^{3x}$

c) $u_1 = -xe^{3x}, u_2 = xe^{3x} - e^{3x}$

d) $u_1 = -xe^{3x} + \frac{1}{3}e^{3x}, u_2 = e^{3x}$

Section 18.3

_____ **1.** A series circuit has a resistor with $R = 15\Omega$, an inductor with $L = 2$, a capacitor with $C = 0.005$ F and a 6 volt battery. An equation for the charge Q at time t is:

a) $2Q'' + 200Q' + 15Q = 6$

b) $2Q'' + 6Q' + 15Q = 200$

c) $2Q'' + 15Q' + 200Q = 6$

d) $2Q'' + 15Q' + 6Q = 200$

_____ **2.** A spring with a 10-kg mass is immersed in a fluid with damping constant 60. To maintain the spring stretched 3 m beyond its equilibrium requires a force of 150 N. The general solution to the equation modeling the motion is:

a) $x = c_1 e^t + c_2 e^{5t}$

b) $x = c_1 e^{-t} + c_2 e^{-5t}$

c) $x = e^{-t}(c_1 \cos 5t + c_2 \sin 5t)$

d) $x = e^{-5t}(c_1 \cos t + c_2 \sin t)$

Section 18.4

____ 1. Find the recursive relation among the coefficients of $y = \sum\limits_{n=0}^{\infty} c_n x^n$ that is the series solution to $y'' = 2y$.

a) $c_n = \dfrac{2c_{n-2}}{n(n-1)}$

b) $c_n = \dfrac{2^n c_{n-2}}{n(n-1)}$

c) $c_n = \dfrac{2c_{n-1}}{n-1}$

d) $c_n = \dfrac{2^n c_{n-1}}{n-1}$

____ 2. The series solution to $y' = xy$ is:

a) $y = \sum\limits_{n=0}^{\infty} \dfrac{c_0}{2^n n!} x^{2n}$

b) $y = \sum\limits_{n=0}^{\infty} \dfrac{c_0}{n!} x^{2n}$

c) $y = \sum\limits_{n=0}^{\infty} \dfrac{2^n c_0}{n!} x^{2n}$

d) $y = \sum\limits_{n=0}^{\infty} \dfrac{n! c_0}{2^n} x^{2n}$

Answers

Chapter 11

Section 11.1

1. C
2. D
3. A

Section 11.2

1. D
2. C
3. B
4. B
5. A
6. B

Section 11.3

1. D
2. C
3. B
4. B
5. B

Section 11.4

1. A
2. D
3. C

Section 11.5

1. C
2. A

Section 11.6

1. C
2. B
3. D
4. C

Chapter 12

Section 12.1

1. B
2. Always
3. Sometimes
4. Always
5. False
6. A
7. D

Section 12.2

1. False
2. True
3. B
4. False
5. C
6. True

Section 12.3

1. C
2. False
3. No
4. Yes
5. B

Section 12.4

1. Never
2. Yes
3. No
4. Yes
5. C

Section 12.5

1. Yes
2. Yes
3. B
4. False

Section 12.6

1. False
2. True
3. True
4. A
5. B
6. True

Section 12.7

1. True
2. False
3. True
4. True

Section 12.8

1. Sometimes
2. True
3. C
4. B
5. C

Section 12.9

1. C
2. C
3. C

Section 12.10

1. C
2. C
3. A
4. True
5. B

Section 12.11

1. B
2. A
3. D

Section 12.12

1. D
2. B
3. B

Chapter 13

Section 13.1

1. A
2. C
3. D
4. D
5. D

Section 13.2

1. B
2. C
3. D
4. False
5. C
6. D

Section 13.3

1. B
2. A
3. B
4. B

Section 13.4

1. D
2. False
3. B
4. D
5. C

Section 13.5

1. D
2. True
3. A
4. D
5. C

Section 13.6

1. B
2. D
3. C
4. B
5. B
6. True

Section 13.7

1. A
2. D
3. D
4. D
5. A

Answers

Chapter 14

Section 14.1

1. A
2. B
3. B

Section 14.2

1. A
2. False
3. True
4. C

Section 14.3

1. A
2. D
3. A
4. True
5. A

Section 14.4

1. B
2. C
3. C
4. A

Chapter 15

Section 15.1

1. C
2. B
3. B
4. C

Section 15.2

1. True
2. A
3. D
4. D
5. False

Section 15.3

1. B
2. B
3. A
4. Sometimes
5. D

Section 15.4

1. False
2. A
3. C
4. C

Section 15.5

1. D
2. A
3. A
4. C

Section 15.6

1. C
2. True
3. B
4. C
5. D

Section 15.7

1. A
2. B
3. B
4. A
5. False

Section 15.8

1. A
2. A

Chapter 16

Section 16.1

1. C
2. D
3. Sometimes

Section 16.2

1. B
2. False
3. B

Section 16.3

1. D
2. False
3. B
4. B

Section 16.4

1. B
2. D

Section 16.5

1. A
2. C
3. No

Section 16.6

1. C

Section 16.7

1. False
2. A
3. A
4. B

Section 16.8

1. A
2. B

Section 16.9

1. D
2. B

Chapter 17

Section 17.1

1. D
2. C
3. B

Section 17.2

1. B
2. True
3. A
4. False
5. A

Section 17.3

1. D
2. B
3. False
4. A

Section 17.4

1. False
2. A
3. D

Section 17.5

1. B
2. True
3. C
4. True
5. C

Section 17.6

1. C
2. C
3. C

Section 17.7

1. C
2. B

Section 17.8

1. B
2. A

Section 17.9

1. D

Chapter 18

Section 18.1

1. True
2. C
3. B
4. False

Answers

Section 18.2

1. A
2. D
3. D

Section 18.3

1. C
2. B

Section 18.4

1. A
2. A